Kicking the Tires?

T0093627

The People's Liberation Army's
Approach to Maintenance Management

JOSLYN FLEMING, CRISTINA L. GARAFOLA,
ELISA YOSHIARA, SALE LILLY, ALEXIS DALE-HUANG

Prepared for the Office of Net Assessment
Approved for public release; distribution is unlimited

NATIONAL DEFENSE RESEARCH INSTITUTE

For more information on this publication, visit **www.rand.org/t/RRA1995-1**.

About RAND

The RAND Corporation is a research organization that develops solutions to public policy challenges to help make communities throughout the world safer and more secure, healthier and more prosperous. RAND is nonprofit, nonpartisan, and committed to the public interest. To learn more about RAND, visit www.rand.org.

Research Integrity

Our mission to help improve policy and decisionmaking through research and analysis is enabled through our core values of quality and objectivity and our unwavering commitment to the highest level of integrity and ethical behavior. To help ensure our research and analysis are rigorous, objective, and nonpartisan, we subject our research publications to a robust and exacting quality-assurance process; avoid both the appearance and reality of financial and other conflicts of interest through staff training, project screening, and a policy of mandatory disclosure; and pursue transparency in our research engagements through our commitment to the open publication of our research findings and recommendations, disclosure of the source of funding of published research, and policies to ensure intellectual independence. For more information, visit www.rand.org/about/principles.

RAND's publications do not necessarily reflect the opinions of its research clients and sponsors.

Published by the RAND Corporation, Santa Monica, Calif.
© 2023 RAND Corporation
RAND® is a registered trademark.

Library of Congress Cataloging-in-Publication Data is available for this publication.

ISBN: 978-1-9774-1267-6

Cover image by Imago/Alamy.

About This Report

The literature on the People's Liberation Army's (PLA's) logistics capabilities is sparse and offers little understanding of subfunctions of logistics, such as maintenance. However, the PLA's growing ability to project and sustain power will depend on its logistics capabilities, systems, and processes. Understanding the PLA's approach to maintenance management is essential in assessing the PLA's ability to sustain a modernized force. This report provides an overview of the PLA's approach to maintenance to inform a broader understanding of how the PLA plans to operate and sustain its forces.

The research reported here was completed in May 2023 and underwent security review with the sponsor and the Defense Office of Prepublication and Security Review before public release.

RAND National Security Research Division

This research was sponsored by the Office of Net Assessment and conducted within the Acquisition and Technology Policy Program of the RAND National Security Research Division (NSRD), which operates the National Defense Research Institute (NDRI), a federally funded research and development center sponsored by the Office of the Secretary of Defense, the Joint Staff, the Unified Combatant Commands, the Navy, the Marine Corps, the defense agencies, and the defense intelligence enterprise.

For more information on the RAND Acquisition and Technology Policy Program, see www.rand.org/nsrd/atp or contact the director (contact information is provided on the webpage).

Acknowledgments

This report would not have been possible without significant contributions from many individuals. We thank Andrew May and Denise Der, of the Office of Net Assessment, for sponsoring this research. Stephan Pikner, Erin Richter, Joshua Arostegui, and J.R. Sessions provided early insights as we framed the problem. We thank our RAND Corporation colleagues Caitlin Lee, Christopher Mouton, Yun Kang, Daniel Romano, James Williams, Bradley Martin, Chad Ohlandt, Michael Kennedy, Nathan Beauchamp-Mustafaga, and John Drew for providing helpful advice and feedback. We thank Carol Evans, George Shatzer, and Roger Cliff of the Strategic Studies Institute for inviting us to attend a conference on PLA sustainment at the U.S. Army War College on March 31–April 2, 2022. Sandra Petitjean brought our findings to life with creative visualizations; Melissa Parmelee refined our prose. We are very grateful to the reviewers of this report, Joel Wuthnow and Bonnie Triezenberg, for their thoughtful comments. Additional individuals and organizations who wish to remain anonymous shared their knowledge with enthusiasm. We thank them greatly for helping bring this research to life.

Summary

The literature on the People's Liberation Army's (PLA's) logistics capabilities is sparse and offers little understanding of subfunctions of logistics, such as maintenance. However, China's growing ability to project and sustain power will depend on its logistics capabilities, systems, and processes. Understanding the PLA's approach to maintenance management is essential for assessing the PLA's ability to sustain a modernized force. This report provides an overview of the PLA's approach to maintenance to inform a broader understanding of how the PLA plans to operate and sustain its forces.

Issue

An assessment of PLA progress in logistics capabilities requires an understanding of the PLA's maintenance apparatus and how that apparatus operates within the PLA's system. We approached this research by examining the following research questions:

- What is the PLA's approach to maintenance?
- How is it controlled and managed?
- How does the PLA's organizational culture affect maintenance?

The PLA has undergone significant reforms over the past few decades, many in logistics. These reforms indicate that the PLA is a logistics force in transition, but it is imperative to understand the objectives of this transition and the PLA's progress toward meeting those objectives. Key areas of interest that we seek to inform with our research include whether the PLA can maintain and sustain its force, particularly considering its growing inventories of modern and advanced power projection systems, and how the PLA's organizational culture of maintenance can provide insight into PLA operations.

Approach

We conducted foundational, exploratory research focused on maintenance for the PLA that we specifically scoped to the PLA Army and PLA Air Force. We created a framework for assessing the operational characteristics and organizational cultures of logistics and maintenance systems and then applied that framework to the PLA to glean insights. To examine the PLA's approach to logistics and maintenance, we drew upon a variety of sources, including PLA professional military education texts, assessments of previous wars fought by the PLA, PLA observations of modern wars, Chinese-language articles published by Chinese state media and military sources, and select English-language secondary sources that have assessed PLA logistics and maintenance practices. Finally, we explored maintenance issues with Chinese weapons and platforms that are fielded by the PLA and exported abroad by comparing Chinese order-of-battle information, arms transfer data, and equipment, which are discussed in PLA articles on maintenance.

Key Findings

Our research indicated that the PLA is a maintenance force in transition and produced the following key findings:

- The PLA has historically viewed maintenance as separate from its logistics functions—a trend that continues today. Therefore, to date, reforms to the Joint Logistics Support Force (JLSF) have little applicability to maintenance functions.
- Primary drivers of logistics reforms include assessments of logistics failures that the PLA experienced in previous conflicts, observations of technological advances in Western militaries, and concerns related to the PLA's lack of combat experience.
- The PLA recognizes the need for a skilled maintainer force to meet the requirements of rapidly modernizing and sophisticated systems.
- One of the critical weaknesses in the PLA's maintenance system is the lack of a professionalized maintenance force. The PLA has made some progress by enhancing the skill level of the force through changes to non-commissioned officer (NCO) training and imbuing NCOs with

more responsibility. However, it is difficult to overcome the large differentiation in skill level between junior and skilled maintainers, particularly when the organizational culture does not prioritize innovation and knowledge-sharing.

- The PLA has prioritized improving its self-identified maintenance weaknesses by instituting process reforms and compensating for low skill levels by developing technological solutions. These solutions mitigate some of the gaps in PLA maintenance proficiency; process reforms are the most promising. However, the fast-paced evolution of weapon system technology might outpace some technology-based compensations.
- An excursion analysis of systems both fielded to the PLA and exported by China identified issues with unmanned aerial vehicle repair parts supply chains that affected system availability.

Recommendations

In this report, we describe the PLA's maintenance practices and the potential relationship between these practices and operations. Because of the exploratory nature of our research, our primary recommendation is to expand it. We have identified the following six areas in which to do so:

- Broaden the scope of the research by assessing maintenance support of large-scale exercises, service-specific exercises, and overseas exercise deployments, among other areas.
- Investigate other services and forces, such as the PLA Navy, which might have similarities to the PLA Army and PLA Air Force but face specific challenges. Examining other branches or units that have been deemed to be less prestigious than combined arms brigades and combat aviation, such as surface-to-air missile forces, might also offer useful insights.
- Study maintenance-related agreements for overseas deliveries of new systems, which might help illuminate the maintenance practices and challenges that are relevant for maintenance forces as China continues to grow the volume, value, and sophistication of its weapon exports.

- Investigate how maintenance interacts with other functions of logistics, such as supply and transportation, that have been affected by the restructuring under the JLSF.
- Understand how maintenance functions directly affect PLA operations.
- Identify and assess the PLA's likely maintenance requirements and challenges for a Taiwan campaign, building on PLA lessons learned from Russia's invasion of Ukraine.

Contents

Figures and Tables

Figures

Tables

CHAPTER 1

Introduction

Logistics is a critical warfighting function that enables military organizations to achieve objectives and complete missions. Understanding the logistics systems of an adversary can provide insight into that adversary's strategic framing and operational approaches, which lays the foundation for assessing the system's overall efficacy.[1] The way in which a military organization plans for and conducts logistics can provide critical insights into how capably the military can achieve its operational objectives. For instance, analysis of Russian logistics capabilities that predates the 2022 invasion of Ukraine indicated that logistics force structure challenges would affect the Russian military's ability to conduct operations on the same timelines that played out in wargames.[2] This assessment explored the challenges that affect the tempo of a potential conflict and provided insights that bore out when the conflict began. The study of logistics systems therefore provides context to how capably a military might perform in actual conflict. Identification of logistics weaknesses could also indicate how a competitor might compensate operationally to account for those deficiencies.

---

[1] James S. Powell, *Taking a Look Under the Hood: The October War and What Maintenance Approaches Reveal about Military Operations*, Institute of Land Warfare, Association of the United States Army, August 2019.

[2] Alex Vershinin, "Feeding the Bear: A Closer Look at Russian Army Logistics and the Fait Accompli," *War on the Rocks*, November 23, 2021.

A military's logistics system can be assessed across any of the seven core functions of logistics.[3] In this report, we focus on maintenance.[4] According to U.S. joint doctrine, maintenance is a subfunction of logistics that enhances the reliability and readiness of systems, enabling freedom of action and affecting the operational readiness of military units.[5] Studying competitors' maintenance capabilities can provide insights into the adaptation patterns, capacity for certain operations, and sustainability of their organizations.[6] Additionally, because of the technical proficiency that is required to maintain rapidly advancing systems, poor maintenance can be one indicator of overall technical skill within an organization. Poor maintenance levels can affect the ability of a unit to operate its systems to their maximum capability. Units might either compensate for poor maintenance through changes to operational design or invest in improvements to directly fix perceived issues. However, change through compensation or improvement might be difficult to implement if organizational culture is not conducive to reform efforts.

Assumptions are often made about the capability of the People's Liberation Army (PLA) to carry out and sustain high-end operations. However, assessments of the PLA's logistics system and its impact on operational capabilities are often lacking in that discussion.[7] A review of literature on

---

[3] According to Joint Publication 4-0, *Joint Logistics*, the seven core logistics functions are "deployment and distribution, supply, maintenance, logistics services, operational contract support (OCS), engineering, and joint health services" (Joint Chiefs of Staff, *Joint Logistics*, Joint Publication 4-0, February 4, 2019, change 1, May 8, 2019, p. x).

[4] We note that the focus of this report is on maintenance, a core function of logistics. However, often in the literature—both the literature that assesses PLA approaches and the organizational literature that we used for our framework—there were few specific mentions of maintenance and a broader focus on logistics. We analyzed the literature for aspects of logistics that are most closely related to maintenance; throughout this report, we occasionally reference logistics more broadly.

[5] Joint Chiefs of Staff, 2019, p. xi.

[6] Powell, 2019, p. 3.

[7] For example, recent public analyses supporting unclassified wargames or scenario development have assumed that the PLA can execute the needed logistics capabilities to launch and sustain high-end joint operations, such as large-scale amphibious operations to invade Taiwan. See Mark F. Cancian, Matthew Cancian, and Eric Heginbo-

2

China's evolving military capabilities indicates that little attention has been given to the topic of the PLA's logistics capabilities, particularly understanding and analyzing the PLA's approach to some key subfunctions of logistics, such as maintenance.[8] Recent analyses have focused on logistics capabilities from an overall delivery focus as the PLA has undertaken reforms over the past two decades that have been aimed at modernizing its military logistics systems to improve support to joint operations and force projection.[9] Functional areas of focus for PLA logistics capabilities include investment in information technologies to support developing precision logistics and logistics cooperation between civilian and military entities. Additionally, the PLA has worked to centralize some elements of logistics under a joint organization, the Joint Logistics Support Force (JLSF). Overall, these

---

tham, *The First Battle of the Next War: Wargaming a Chinese Invasion of Taiwan*, Center for Strategic International Studies, January 2023; David Lague and Maryanne Murray, "T-Day: The Battle for Taiwan," Reuters, November 5, 2021; and Stacie Pettyjohn, Becca Wasser, and Chris Dougherty, *Dangerous Straits: Wargaming a Future Conflict over Taiwan*, Center for a New American Security, June 2022. Cancian, Cancian, and Heginbotham notes some dynamics for PLA supply lines but does not focus on other elements of PLA logistics.

[8] Two key exceptions include a 2022 conference on PLA logistics topics and recent work on cross–Taiwan Strait logistics and mobilization. See George R. Shatzer and Roger D. Cliff, eds., *PLA Logistics and Sustainment: PLA Conference 2022*, US Army War College Press, 2023. On cross–Taiwan Strait logistics and mobilization, see Chung Chieh and Andrew N. D. Yang, "Crossing the Strait: Recent Trends in PLA 'Strategic Delivery' Capabilities," in Joel Wuthnow, Arthur Ding, Phillip C. Saunders, Andrew Scobell, and Andrew N.D. Yang, eds., *The PLA Beyond Borders: Chinese Military Operations in Regional and Global Context*, National Defense University Press, 2021; Chieh Chung, "PLA Logistics and Mobilization Capacity in a Taiwan Invasion," in Joel Wuthnow, Derek Grossman, Phillip C. Saunders, Andrew Scobell, and Andrew N. D. Yang, eds., *Crossing the Strait: China's Military Prepares for War with Taiwan*, National Defense University Press, 2022; Lonnie D. Henley, *Civilian Shipping and Maritime Militia: The Logistics Backbone of a Taiwan Invasion*, China Maritime Studies Institute, May 2022; and Kevin McCauley, *Logistics Support for a Cross-Strait Invasion: The View from Beijing*, China Maritime Studies Institute, 2022.

[9] LeighAnn Luce and Erin Richter, "Handling Logistics in a Reformed PLA: The Long March Toward Joint Logistics," in Phillip C. Saunders, Arthur S. Ding, Andrew Scobell, Andrew N. D. Yang, and Joel Wuthnow, eds., *Chairman Xi Remakes the PLA: Assessing Chinese Military Reforms*, National Defense University Press, 2019; Joel Wuthnow, "A New Era for Chinese Military Logistics," *Asian Security*, Vol. 17, No. 3, September–December 2021.

reforms point to the PLA as a logistics force in transition, but it is imperative to understand the goal of that transition. Can the PLA sustain and maintain its forces, particularly considering its growing inventories of modern and advanced power projection systems?

These questions are particularly interesting because the PLA is rapidly modernizing its military systems, which requires sophisticated mainte-nance management practices to keep pace and maintain fleets of much more complex systems. The fielding of new systems requires increased main-tenance functions and skills compared with the fielding of older systems, often because more advanced technology requires more maintenance than older technology. Common approaches to modernizing militaries include the need for increased technical skill levels for maintenance personnel, the ability to leverage cooperation between civilian and military maintenance providers, the maintenance of a multitude of systems that lack commonal-ity, and the transition from disposable systems to those that require invest-ment in maintenance capabilities. An assessment of PLA progress in these capabilities requires an understanding of the PLA's maintenance apparatus and how it operates within the larger system. What is the PLA's approach to maintenance? How is maintenance controlled and managed? How does the PLA's organizational culture affect maintenance?

As we explore in this report, maintenance is more closely tied with engi-neering functions than it is with logistics functions in the PLA. This dynamic raises the question of whether the PLA even considers maintenance to be a subfunction of logistics. Maintenance also seems to be uniquely immune to the recent reforms aimed at centralizing logistics under a joint organization (the JLSF). While such subfunctions as transportation have been profoundly changed under these organizing principles, maintenance structures remain, for the most part, similar to prereform structures. We find that these logis-tics reforms inform maintenance approaches but do not directly change how maintenance is controlled and managed.

Research Approach

In the sections that follow, we outline our research approach.

Exploratory Research

We conducted foundational, exploratory research focused on the logistics subfunction of maintenance. Our assessments of the PLA's approach drew on decades of RAND expertise on logistics and maintenance and were inspired by RAND work that assessed the Soviet Union during the Cold War. Inspired by previous RAND work, such as research by Andris Trapans, we were mindful to not mirror image U.S. perspectives.[10]

Methodology

We began our research by developing a framework for focusing our assessment of logistics and, by extension, maintenance systems on operational characteristics and aspects of organizational culture that are key indicators of how well an organization performs logistics and maintenance functions. The framework was built by conducting a literature review of sources that focus on organizational theory, military studies, and historic case analyses that referenced logistics-related practices. There are limited sources that are specific to military logistics , so we expanded the scope to include logistics factors that affected similar types of civilian organizations. This framework is further explained in Chapter 3 and Appendix A. We used this framework for our assessment of PLA maintenance approaches in Chapter 4.

To examine the PLA's approach to logistics and maintenance, summarized in Chapter 2, we drew upon a variety of sources, including PLA professional military education texts, assessments of previous wars in which the PLA has fought, and PLA observations of modern wars. To understand maintenance practices, we collected Chinese-language articles published by *PLA Daily*; other Chinese state media and military sources, such as *People's Daily*; PLA service and theater command (TC) news sites; provincial and regional news outlets' and select nonauthoritative Chinese media sources. We discuss the benefits and limitations of these sources in Chapter 4.

---

[10] The work of Trapans was fundamental in developing our research. Please see Andris Trapans, *Organizational Maintenance in the Soviet Air Force*, RAND Corporation, RM-4382-PR, 1965; and Andris Trapans, *Logistics in Recent Soviet Military Writing*, RAND Corporation, RM-5602-PR, 1966.

We additionally reviewed select English-language secondary sources that have assessed PLA logistics and maintenance practices, typically by Western scholars who study the PLA. Finally, we explored maintenance issues with Chinese weapons and platforms that are fielded by the PLA and exported abroad by China by comparing Chinese order-of-battle information, arms transfer data, and equipment that is discussed in PLA articles on maintenance.

Scope

We focused our research on understanding units and forces based within mainland China, along with the systems and organizations that shape their maintenance practices. We adopted this scope for the following three reasons:

- Information on maintenance practices and trends within two services—the PLA Army (PLAA) and PLA Air Force (PLAAF)—was more readily available than information on such services as the PLA Rocket Force (PLARF).[11]
- In our initial review, it appeared that PLA Navy (PLAN) maintenance practices are not especially similar to PLAA and PLAN maintenance dynamics. Some differences that we observed include more-predominant leveraging of military-civil fusion (MCF) and more-substantive differences between maintenance that is conducted in port and maintenance conducted at sea. Therefore, we opted to remove PLAN maintenance from the scope of our research, although we acknowledge that it would be a promising area for future research.
- Although the PLA's ability to improve its capabilities for out-of-area expeditionary operations is of keen interest for analysts and policymakers, existing research on this topic has identified a host of challenges that the PLA has yet to overcome.[12]

---

[11] This could be because of PLA sensitivities around discussing details regarding missile force capabilities in media sources.

[12] See, for example, Cristina L. Garafola and Timothy R. Heath, *The Chinese Air Force's First Steps Toward Becoming an Expeditionary Air Force*, RAND Corporation, RR-2056-AF, 2017; Cristina L. Garafola, Timothy R. Heath, Christian Curriden, Meagan L. Smith, Derek Grossman, Nathan Chandler, and Stephen Watts, *The Peo-*

How This Report Is Organized

In this report, we explore the PLA's maintenance practices and analyze trends in its approach that provide insight into its operational capabilities. In Chapter 2, we assess how the PLA logistics and maintenance apparatus has evolved from an organizational perspective. We first outline previous reforms and the impetus for those reforms, then characterize the current system. In Chapter 3, we provide a framework for the exploration of maintenance practices as they interact with organizational culture. We present logistics success factors that can inform the analysis of trends in maintenance approaches. In Chapter 4, we explore the operational characteristics, organizational culture, and ability of the PLA's maintenance apparatus to adapt by applying the framework provided in Chapter 3. We present conclusions, implications, and areas for further research in Chapter 5. Appendix A explains additional information about a subset of the logistics success factors that are introduced in Chapter 3. Appendix B lists key terms of reference used by Chinese sources to discuss logistics and maintenance concepts.

---

ple's Liberation Army's Search for Overseas Basing and Access: A Framework to Assess Potential Host Nations, RAND Corporation, RR-A1496-2, 2022; Kristen Gunness, "The Dawn of an Expeditionary PLA?" in Nadège Rolland, ed., *Securing the Belt and Road Initiative: China's Evolving Military Engagement Along the Silk Road,* National Bureau of Asian Research, September 2019; Luce and Richter, 2019; Wuthnow, 2021; Joel Wuthnow, Arthur S. Ding, Phillip C. Saunders, Andrew Scobell, and Andrew N. D. Yang, eds., *The PLA Beyond Borders: Chinese Military Operations in Regional and Global Context,* National Defense University Press, 2021; and Joel Wuthnow, Phillip C. Saunders, and Ian Burns McCaslin, "PLA Overseas Operations in 2035: Inching Toward a Global Combat Capability," *Strategic Forum,* No. 309, National Defense University Press, May 2021.

CHAPTER 2

The Evolution of the People's Liberation Army's Approach to Logistics and Maintenance

As both a student of its own historical combat experience and a keen observer of the crucial role that logistics and maintenance have played in modern warfare, the PLA has absorbed many lessons and sought to adapt those lessons to its own system. The PLA views logistics and maintenance as integral components of its approach to *systems-of-systems thinking* and *systems confrontation*—the foundational concepts by which the PLA seeks to organize and execute modern warfare to prevail against opponents.[1]

In this chapter, we first examine the PLA's efforts to reform its logistics systems, which predate recent military reforms in the era of General Secretary Xi Jinping and are in response to challenges that Chinese analysts had previously identified in the PLA's logistics capability, including challenges regarding maintenance. Maintenance does not appear to be considered a subfunction of logistics in the PLA like it is in the U.S. military. However, understanding logistics challenges provides insight into those areas that directly affect maintenance and additional context regarding the broader organizational culture that shapes the PLA's maintenance practices. We then summarize the position of maintenance within the PLA's organiza-

---

[1] We thank our reviewer, Joel Wuthnow, for raising this point. For more information on the role of the support system within the PLA's system-of-systems thinking, see Jeffrey Engstrom, *Systems Confrontation and System Destruction Warfare: How the Chinese People's Liberation Army Seeks to Wage Modern Warfare,* RAND Corporation, RR-1708-OSD, 2018.

tional structure, with an emphasis on the post–2015 and 2016 PLA reform era, and offer some insights into the relationship between maintenance and PLA logistics from an organizational perspective. We conclude with a broad mapping of modern-day maintenance activities that have been undertaken in two PLA services, the PLAA and PLAAF, which sets the stage for our analysis of PLA maintenance practices and management in Chapter 4.

Early Efforts to Reform People's Liberation Army Logistics

Many modern-day assessments view the large-scale PLA-wide reforms of the 2015 to 2016 period as resulting in the most significant transformation in the PLA's modern history.[2] However, large-scale efforts to reform PLA's logistics systems and processes predate Xi's announcement of major PLA reforms by nearly two decades.[3] Three key drivers of logistics reform included logistics failures during the Chinese military's participation in previous conflicts, observations of logistics and maintenance advances in Western militaries, and a desire to realize greater efficiencies within the logistics system.[4]

Historical Challenges and Drivers for Reforms

PLA officials recognize the critical role of logistics and maintenance in enabling warfighting operations. For example, the 2001 version of the *Science of Military Strategy* (SMS) states that "modern warfare is not only about going to war in a military sense, it is about going to economic war. This puts

---

[2]   See, for example, Office of the Secretary of Defense, *Annual Report to Congress: Military and Security Developments Involving the People's Republic of China 2019*, Department of Defense, May 2019, p. ii.

[3]   For a Chinese perspective on China's earlier attempts to reform the PLA logistics structure prior to 1998, see Nong Qinghua [农清华], "Reform of PLA Logistics Support System in the Past 40 Years" [人民解放军后勤保障体制改革攻坚40年], *Military History* [军事历史], No. 1, 2019.

[4]   Roger Cliff, *China's Military Power: Assessing Current and Future Capabilities*, Cambridge University Press, 2015, p. 141; Luce and Richter, 2019, p. 261.

forward even higher requirements for strategic logistics support work."[5] On maintenance, SMS 2001 notes that "the experiences of past wars have demonstrated the fact that the quality of technical maintenance has direct impact on the perfect performance of the equipment and weapons systems, and furthermore, has bearing on the success or failure of strategic weapons."[6] The first driver informing this recognition is demonstrated by China's previous military experiences, which have forced PLA leaders to critically evaluate the ability of logistics forces to support future military operations.

Logistics Failures During the People's Liberation Army's Participation in Previous Conflicts

Approaches adopted by the PLA during World War II and the Chinese Civil War against the Kuomintang forces, such as self-sufficiency, sometimes compensated for the lack of a production base and logistics system through the capture of weapons and supplies from enemy forces to construct ordnance factories and the seizure of food from local peasants.[7] Chinese Communist Party (CCP) leaders recognized that requisitioning food from the public was particularly unpopular and sought to reduce demands on the peasant population by growing crops and raising animals. Mao Zedong

---

[5] Peng Guangqian [彭光谦] and Yao Youzhi [姚有志], eds., *Science of Military Strategy* [战略学], Academy of Military Science Press [军事科学出版社], 2001, p. 376. Notably, SMS 2001 describes strategic logistics support as a separate category from strategic equipment support, which the text describes as including maintenance. A more recent edition in this series, SMS 2013, likewise describes logistics and equipment support separately. SMS 2020 (the most recent volume on this topic but, unlike the other volumes, authored by National Defense University [NDU]) also separates the two categories. See Peng and Yao, 2001, p. 378; Shou Xiaosong [寿晓松], ed., *Science of Military Strategy* [战略学], Academy of Military Science Press [军事科学院], 2013, p. 266; and Xiao Tianliang [肖天亮], ed., *Science of Military Strategy* [战略学], National Defense University Press [国防大学出版社], 2020, pp. 83, 205.

[6] Peng and Yao, 2001, p. 358.

[7] Gary J. Bjorge, *Moving the Enemy: Operational Art in the Chinese PLA's Huai Hai Campaign*, Combat Studies Institute Press, 2004, pp. 37, 52. See also John J. Tkacik, Jr., "From Surprise to Stalemate: What the People's Liberation Army Learned from the Korean War—A Half-Century Later," in Laurie Burkitt, Andrew Scobell, and Larry M. Wortzel, eds., *The Lessons of History: The Chinese People's Liberation Army at 75*, Strategic Studies Institute Press, 2003, p. 302.

described this approach as having "an army for fighting as well as an army for labor. For fighting we have the Eighth Route and New Fourth Armies; but even they do a dual job [of] warfare and production."[8] Self-sufficiency efforts, particularly during the Cultural Revolution, would contribute to the PLA's ineffectiveness in later conflicts because of the time and effort troops were spending on self-sufficiency initiatives rather than on building and sustaining operational capability.

After the CCP defeated the Kuomintang and founded the People's Republic of China in 1949, however, the PLA continued to face a variety of logistics-related challenges in the Korean War and Sino-Vietnamese War. A memoir by Hong Xuezhi, the logistics department head for the Chinese People's Volunteer Force (CPVF) during the Korean War, notes some comparative strengths in China's logistics system but also recognizes key weaknesses. Compared with transoceanic supply lines for U.S. and other United Nations Command forces, the Chinese military benefited from close rear areas and lower requirements for fuel and munitions on a per-platform basis for key weapons, such as large artillery pieces, and vehicles.[9] However, Hong also documents major logistics challenges. These challenges included an understaffed regional logistics command; repeated American air raids on materiel storage stations, rail lines, and road networks that were crucial to CPVF supply truck movements; inexperienced logistics officers and truck drivers; and rapidly lengthening supply lines.[10] The CPVF's efforts to procure supplies locally by requesting permission from local Korean authorities to borrow food and grain from the populace also failed in conflict zones that were already devastated by the war.[11] Other sources recount how shortages of rations, winter uniforms, and ammunition on the front lines contributed

---

[8] As quoted in Dennis J. Blasko, "PLA Ground Forces Lessons Learned: Experience and Theory," in Laurie Burkitt, Andrew Scobell, and Larry M. Wortzel, eds., *The Lessons of History: The Chinese People's Liberation Army at 75*, Strategic Studies Institute, July 2003, p. 64.

[9] Hong Xuezhi, "The CPVF's Combat and Logistics," in Xiaobing Li, Allan R. Millett, and Bin Yu, eds., *Mao's Generals Remember Korea*, University Press of Kansas, 2001, p. 116. The cited chapter is adapted from Hong's memoir.

[10] Hong, 2001, pp. 123–130. See also Tkacik, 2003, p. 305.

[11] Hong, 2001, p. 127.

to tens of thousands of deaths in winter and left troops resorting to bayonet charges during offensive operations.[12] One assessment concludes that the military "outr[unning] its supply lines . . . result[ed] in a stalemate on the 38th parallel." Moreover, the "PLA logistics supply was relentlessly pounded by U.S. airpower and Chinese forces at the front suffered accordingly. That original 'trial by fire' made a deep impression upon the PLA regarding the importance of military logistics."[13] Like logistics, PLA maintenance at this time was most likely structured around the assumption that rear-area support would be located nearby. Drawn-out supply lines would affect the ability of maintainers to provide maintenance support at the front because the provision of repair parts would be delayed.

More logistics challenges hampered the PLA's performance during the 1979 Sino-Vietnamese War. Severe food and water shortages affected fighting units, and shortfalls could not be offset through local requisitioning because the fighting was taking place in hostile territory.[14] Other deficiencies included failing to supply sufficient rations to units prior to the invasion and possible loss of some rations during the fighting because of troop inexperience, lack of discipline, or stress.[15] One scholar found that, despite PLA logistics support taking months to prepare, "China's military logistics system proved inadequate to sustain offensive military operations in Vietnam in 1979. Consequently, China's military forces were forced to fall back to positional warfare . . . along China's southern border during operations in Vietnam."[16] A 1980 assessment by the Chinese General Logistics Department (GLD) on problems with the logistics support provided during the war concluded that because "the support system has a great impact on the ben-

---

[12] Tkacik, 2003, pp. 302–303.

[13] Lyle J. Goldstein, "America Cannot Ignore China's Military Logistics Modernization," *National Interest*, October 11, 2021.

[14] Edward C. O'Dowd, *Chinese Military Strategy in the Third Indochina War: The Last Maoist War*, Routledge, 2007, pp. 70–71.

[15] O'Dowd, 2007, p. 71.

[16] Susan M. Puska, "Taming the Hydra: Trends in China's Military Logistics Since 2000," in Roy Kamphausen, David Lai, and Andrew Scobell, eds., *The PLA at Home and Abroad: Assessing the Operational Capabilities of China's Military*, Strategic Studies Institute, 2010, p. 554.

efits [provided], [we] need to make reforms to the support system the main subject of [our] logistics reforms."[17]

Figure 2.1 summarizes the PLA's experience during these three conflicts, as well as logistics- and maintenance-relevant adaptions and struggles. The PLA made efforts to compensate for a lack of robust logistics processes and durable supply lines and reduce overall logistics burdens through self-sufficiency measures. However, repeated challenges with sustaining a comprehensive logistics system detracted from the military's capability-building during peacetime and led to a variety of operational problems during conflict.

Observations of Logistics and Maintenance Advances in Western Militaries

Reforms have also been shaped by a recognition that the PLA's capabilities and readiness were inferior to other militaries' warfighting capabilities. PLA scholars have reviewed both historical and contemporary conflicts to understand logistics successes and other relevant dynamics. Contemporary Chi-

FIGURE 2.1

Culture of Compensation and Drive for Self-Sufficiency Were Key Elements of the People's Liberation Army's Historical Approach to Wartime Logistics and Maintenance

PLA logistics and maintenance experience	Captured weapons and supplies from enemy Seized food from peasants	Faced understaffed regional logistics command, attacks on supply lines Conflict zone unable to provide food supplies	Failed to supply enough rations to units prior to deployment Could not requisition supplies in hostile territory
	World War II and Civil War 1937–1949	**Korean War** 1950–1953	**Sino-Vietnamese War** 1979
Adaptations and struggles	Eighth Route and New Fourth Armies tasked to grow crops and raise animals; self-sufficiency roles later detracted from operational mission	Improved supply line resiliency, but ultimately outran supply lines, resulting in shortages and a stalemate	Despite months of preparation, logistics system unable to sustain offensive operations in Vietnam

[17] Nong, 2019, p. 11.

nese analysis of the U.S. armed forces' successes against Japan during the Pacific War (1941–1945) draws insights about both militaries' logistics performance.[18] Table 2.1 summarizes these insights, noting which campaigns are referenced and which military met with logistics successes or failures.

TABLE 2.1

Contemporary People's Liberation Army Insights on Logistics Success and Failures During the Pacific War

Theme	Successes and Related Failures, If Applicable
Develop robust logistics footprint	Success: Employ vast numbers of support vessels (Okinawa campaign, United States)[a] • Related failure: Using warships for resupply, which was inefficient and reduced combat power (Guadalcanal campaign, Japan)[a] Success: Seize territory that augmented existing supply chains (Okinawa campaign, United States)[a]
Attack opponent's logistics system	Success: Interdict the opponent's resupply and destroy supply lines (Guadalcanal campaign, United States)[a] • Related failure: Focusing on counterforce operations rather than exploiting the other side's logistics-related vulnerabilities (Guadalcanal campaign, Japan)[a]
Protect logistics system	Success: Employ air and naval forces to protect the entire supply chain between the rear area and the frontlines (Guadalcanal campaign, United States)
Leverage multimodal airpower	Success: Leverage land-based airpower, which conferred advantages in range and capacity compared with naval aviation alone (Guadalcanal campaign, United States)[a]
Optimize logistics system efficacy	Success: Carefully manage long-distance supply lines to ensure timely and consistent arrival of goods, ammunition, and support vessels for combat ships (Okinawa campaign, United States) Success: Efficiently concentrate unloading at debarkation points (Okinawa campaign, United States)[a]

SOURCE: Toshi Yoshihara, *Chinese Lessons from the Pacific War: Implications for PLA Warfighting*, Center for Strategic and Budgetary Assessments, January 5, 2023, pp. 49–52, 62–64.

NOTE: The Chinese sources cited by Yoshihara date from 1996 to 2020.

[a] These insights are noted in at least one source dating from 2008 (when the PLA completed a major round of logistics reforms) or earlier, which indicates that these insights were already published when the PLA was planning or implementing reforms.

---

[18] For logistics and other lessons learned by the PLA from the Pacific War, see Yoshihara, 2023.

PLA observations of more-recent conflicts also inform views on modern logistics and maintenance requirements. A 2000 study by the PLA's General Staff Department (GSD) on lessons learned from the Kosovo War identifies logistics and maintenance successes and challenges.[19] Observed U.S. and North Atlantic Treaty Organization (NATO) successes included leveraging pre-positioning near the conflict; employing significant strategic airlift for rapid logistical support; and leveraging a large number of dispersed bases in allied territory for aircraft refueling, rearming, maintenance, and crew support.[20] The GSD authors also note the success of harnessing civilian capacity, including mobilizing civilian aircraft and reserve personnel, purchasing materiel via commercial contracts, and leveraging contractors at air bases.[21]

The GSD authors found that then-Yugoslavia faced NATO strikes on its road networks, rail lines, and bridges that damaged logistics supply lines.[22] Supply shortages because of NATO strikes and an energy embargo ultimately limited Yugoslavia's ability to continue combat operations, even though it leveraged pre-positioned materiel, mobilization of civilians, and denial and deception operations to protect its logistical support.[23] The authors also observed that Yugoslavia had underinvested in both modern military equipment and maintenance of its equipment prior to the war, which affected both military capability and troop morale.[24] PLA NDU authors of a similar study on the Kosovo War concluded that "vigorously developing support equipment and improving support facilities is an urgent task for our military's logistics construction. . . . Establishing an integrated . . . logistics support system is an objective requirement for modern warfare."[25]

---

[19] PLA GSD Military Training Department, *Research into the Kosovo War*, Liberation Army Publishing House, 2000.

[20] PLA GSD Military Training Department, 2000, Chapter 3, Section 5.

[21] PLA GSD Military Training Department, 2000, Chapter 3, Section 5.

[22] PLA GSD Military Training Department, 2000, Chapter 3, Section 3.

[23] PLA GSD Military Training Department, 2000, Chapter 3, Section 5.

[24] PLA GSD Military Training Department, 2000, Chapter 5, Section 2.

[25] Zeng Tingze [曾廷泽] and Pan Dahong [潘大红], "The Enlightenment of Logistic Support from Kosovo War" [从科索沃战争看未来后勤保障], *Military Economic Research* [军事经济研究], 2000, p. 74.

Likewise, in a PLA study on the Afghanistan War, the authors credited the United States with rapidly establishing a long-distance but flexible logistics system that leveraged allies' bases; military and civilian airpower; pre-positioned naval ships; and helicopters, new ground vehicles, and donkeys to adapt to Afghanistan's terrain.[26] The United States also successfully cut off the Taliban's supply lines at Mazar-e-Sharif.[27] The authors additionally credit U.S. forces with standing up significant equipment maintenance forces in the field to maintain good working order of equipment and adapt equipment to harsh local weather conditions.[28] The authors concluded that the PLA should improve rapid repair capabilities, among others, for its counterterrorism forces.[29]

In a third PLA study, on the Iraq War, the analysts focused on the U.S. ability to harness large rear-area logistics bases in friendly territory close to the Iraqi border.[30] The PLA authors noted that the Iraqi forces had some successes attacking U.S. and United Kingdom logistics units, but ultimately those attacks were not significant enough to disrupt the U.S. network supply lines that supported the front line.[31] Following a major disarmament and arms sales restrictions, as well as sanctions stemming from the Gulf War, the Iraqi military was limited to refurbishing or improving existing equipment, and operational aircraft were in short supply leading up to the war.[32] Supply shortages of required repair parts affect how maintenance can be conducted. Militaries might compensate for shortages by cannibalizing from other equipment, but doing so affects overall equipment readiness and is difficult to sustain. The authors also noted that the Iraqi difficulties

[26] PLA GSD Military Training and Arms, *Research into the Afghanistan War*, Liberation Army Press, 2004.

[27] PLA GSD Military Training and Arms, 2004, Chapter 4, Section 2, pp. 9–10.

[28] PLA GSD Military Training and Arms, 2004, Chapter 7, Section 2, pp. 99–100.

[29] PLA GSD Military Training and Arms, 2004, Chapter 9, Section 5.

[30] Wang Yongming, Liu Xiaoli, and Xiao Yunhua, eds., *Research into the Iraq War*, Military Science Press, 2003, Chapter 2, Section 2, p. 34.

[31] Wang, Liu, and Xiao, 2003, Chapter 5, Section 2; Chapter 7, Section 2, pp. 108, 180.

[32] Wang, Liu, and Xiao, 2003, Chapter 2, Section 3, p. 46.

repairing or replacing equipment were significant weaknesses that limited combat effectiveness.[33]

Figure 2.2 summarizes PLA observations from the Pacific War and the three modern wars that were the subject of the studies discussed previously. Contemporary PLA authors writing about the Pacific note the ability of the United States to establish, maintain, and defend logistics supply lines, which they contrasted with Japan's ineffectiveness at doing so along with Japan's failure to target U.S. logistics vulnerabilities when there were opportunities to do so. Taken together, PLA observations regarding the three modern conflicts identify commonalities to successful and unsuccessful approaches. U.S. and coalition forces are credited with protecting vast supply lines, ensuring timely arrival of goods and services, and prioritizing maintenance and repair of highly capable equipment. In contrast, although the opponents in some cases benefited from large-scale mobilization and pre-positioned stocks in organized rear areas, most faced severe challenges from U.S. or coalition strikes on their supply lines. Longstanding weaknesses in maintenance capability further limited these militaries' ability to reconstitute equipment during wartime.

PLA authors recognize that China's lack of combat experience and opportunities to test logistics capabilities since the Sino-Vietnamese War present a challenge for assessing the PLA's modern logistics capabilities. Authoritative Chinese military texts have expressed particular concern over the PLA's lack of combat experience and its effect on maintenance. For example, the PLA NDU's 2017 SMS argues that, in regard to maintenance and support capability for high-technology weapons and equipment, "most of our new weapons and equipment have not been tested in actual combat, and under harsh battlefield conditions may have a higher failure rate." The authors conclude that "it is necessary to further strengthen the construction of equipment maintenance support teams, conduct targeted professional training, and comprehensive drills, and effectively enhance the battlefield repair capabilities of high-tech weapons and equipment."[34]

---

[33] Wang, Liu, and Xiao, 2003, Chapter 7, Section 2, pp. 180–181.

[34] Xiao Tianliang [肖天亮], ed., *Science of Military Strategy* [战略学], National Defense University Publishing House [国防大学出版社], 2017, p. 433.

FIGURE 2.2

People's Liberation Army Observations of Modern Wars Highlight U.S. Successes and Opponent Failures as Lessons Learned for Logistics and Maintenance Modernization

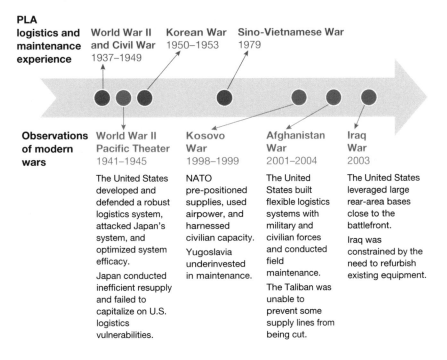

PLA logistics and maintenance experience	World War II and Civil War 1937–1949	Korean War 1950–1953	Sino-Vietnamese War 1979

Observations of modern wars	World War II Pacific Theater 1941–1945	Kosovo War 1998–1999	Afghanistan War 2001–2004	Iraq War 2003
	The United States developed and defended a robust logistics system, attacked Japan's system, and optimized system efficacy. Japan conducted inefficient resupply and failed to capitalize on U.S. logistics vulnerabilities.	NATO pre-positioned supplies, used airpower, and harnessed civilian capacity. Yugoslavia underinvested in maintenance.	The United States built flexible logistics systems with military and civilian forces and conducted field maintenance. The Taliban was unable to prevent some supply lines from being cut.	The United States leveraged large rear-area bases close to the battlefront. Iraq was constrained by the need to refurbish existing equipment.

NOTE: Time spans for Afghanistan and Iraq Wars reflect the years analyzed in PLA studies, not the actual years of those conflicts.

Desires to Realize Greater Efficiencies Within the Logistics System

Emerging from the nadir of operational effectiveness experienced during the Cultural Revolution, the PLA began the 1980s with a gradual withdrawal from civilian roles to refocus on military tasks and missions.[35] With regard to logistics, key desired areas of improvement and reform included

---

[35] Puska, 2010. Thanks to our reviewer, Joel Wuthnow, for highlighting the theme of realizing efficiencies. For more on the PLA's decline in operational capability during this period, see O'Dowd, 2007.

reducing redundancy in logistics organizations' capacity and focus as well as staffing, achieving cost-savings associated with reducing redundancy and a more-efficient allocation of resources, and increasing reliance on civilian entities and the commercial sector. Curbing the corruption that persisted in the logistics system through the 1980s and 1990s was also a goal, given that corruption impeded logistics-specific efficiencies and undermined broader military operational effectiveness.

1998–2008 Logistics Reforms

The PLA's understanding of its logistics shortfalls in historical conflicts, developments in modern warfare, and desires to realize greater efficiencies informed one of the most significant efforts to reform PLA logistics. The reforms were initiated in 1998 as part of a ten-year plan and in support of a broad effort to modernize China's armed forces. Although China attempted to reform the structure of the PLA's logistics forces prior to 1998, Western analysts have argued that the military reforms that began under Jiang Zemin in 1998 were "a necessary precursor to real joint logistics command integration."[36] The PLA's ten-year reform plan resulted in improvements to the structure of China's logistics apparatus, including the establishment of a basic joint logistics structure and the elimination of the largest vertical redundancies.

A prominent objective of the ten-year reform plan was the creation of a joint logistics system in each of China's military regions in 2000, which would end the structure of separate logistics system for the PLAN, PLAAF, and Second Artillery (which was not yet an independent service).[37] Prior to the reforms, each service headquarters in Beijing controlled its own logistics apparatus, while the ground forces were supported by the regional logistics departments under the military region system. This structure led to multiple separate and redundant logistics organizations with minimal overlap for even the most basic materiel. The reforms focused on moving these basic elements of logistics to a more joint system to eliminate redundant stove-

---

[36] Luce and Richter, 2019, p. 262.

[37] Lonnie Henley, "PLA Logistics and Doctrine Reform, 1999-2009," in Susan M. Puska, ed., *People's Liberation Army After Next*, Strategic Studies Institute, August 2000, p. 57.

pipes. Changes included merging support activities common across the services into "new Joint Logistics [Sub-] Departments (JLDs) in each of the military regions' logistics departments," although each service retained a logistics system for specialty materiel and support. The same assessment also notes that

> under the rubric of "network-style zoned supply work," the services are surrendering hospitals, medical supply and maintenance units, fuel stocks and distribution networks, general supply warehouses, vehicle supply and maintenance units, and some general transportation units to the control of the MR [military region] JLD.[38]

Chinese military scholars and officials praised these changes for reducing redundancy and setting conditions for more-integrated joint operations. One scholar affiliated with the NDU noted that this specific change made the PLA "take a big step in the direction of integrating the three armed services and collectivizing logistics support."[39] A deputy minister of the GLD's Propaganda Ministry characterized the change as "breaking through the historical pattern of the three services being self-contained."[40] In 2004, the PLA decided to consolidate the three services' separate logistics systems in the Jinan Military Region within the region's JLD.[41] This consolidation allowed the PLA to further test more-integrated joint capabilities as a precursor to establishing the JLSF.

---

[38] Henley, 2000, p. 58.

[39] Nong, 2019, p. 12.

[40] Zhao Jianwei [赵建伟], "China's Military Logistics Reform Plan Is Clear, and Substantial Progress Has Been Made in Four Aspects" [中国军事后勤变革图渐清晰 四方面获实质性进展], *ChinaNews* [中国新闻网], April 24, 2007.

[41] "A Perspective on China's Military Logistics Transformation: Substantial Steps in Four Areas (2)" [透视中国军事后勤大变革：四方面迈出实质性步伐(2)], China News [中国新闻网], January 12, 2007; Cliff, 2015, p. 148; Xu Jinzhang [胥金章] and Zhang Yuqing [张玉清], "With the Approval of the Central Military Commission, the Jinan Theater Has Carried Out a Pilot Program of Large-Scale Joint Logistics Reform" [经中央军委批准 济南战区进行大联勤改革试点], Xinhua, June 25, 2004.

Another core component of China's ten-year logistics reform plan was the *socialization* (社会化) of logistics services.[42] By the 1980s and 1990s, the drive for greater self-sufficiency within the PLA—which was meant to first reduce the burden on local peasants and then to keep government defense expenditures low—resulted in the development of agriculture; light industry; and other commercial ventures, such as hospitality.[43] Corruption and distraction from key military operational activities, such as training, followed. As the PLA divested of nonmilitary businesses, such as restaurants, kindergartens, and laundry shops, it handed off some of these responsibilities to China's civilian sectors, including transportation, medical, construction, and food services.[44] This change also had a direct impact on basic maintenance functions. For example, the PLA developed contract relationships with civilian vehicle servicing shops for maintenance and military vehicle refueling purposes.[45] The socialization of logistics services allowed China's armed forces to prioritize operational tasks and meant that approximately 200,000 personnel "were no longer on the PLA's payroll."[46]

---

[42] For details on China's "socialization" of logistics responsibilities, see "Circular of the State Council and the Central Military Commission on Issues Concerning Promoting the Socialization of Military Logistics Support" [国务院、中央军委关于推进军队后勤保障社会化有关问题的通知], Central People's Government of the People's Republic of China [中华人民共和国中央人民政府], June 14, 2018. For additional PLA analysis on implementing Jiang Zemin's guidance regarding military-civil integration relevant for logistics and maintenance, see Sun Xiude [孙秀德] and Xu Qi [徐起], "The Direction of Military Logistic Support System Reforms [军队后勤保障体制改革的方向], *Military Economic Research* [军事经济研究], Vol. 7, 1998, pp. 15–22.

[43] Blasko, 2003, p. 64.

[44] Cliff, 2015, p. 148; Luce and Richter, 2019, p. 268. See also Dennis J. Blasko, *The Chinese Army Today: Traditions and Transformation for the 21st Century*, Routledge, 2012a, p. 35.

[45] Blasko, 2012a, p. 35.

[46] Henley, 2000, p. 60; Erin Richter, Leigh Ann Ragland, and Katherine Atha, "General Logistics Department Organizational Reforms: 2000-2012" in Kevin Pollpeter and Kenneth W. Allen, eds., *The PLA as Organization v2.0*, Defense Group Inc., 2012, p. 178.

Continuing Challenges in People's Liberation Army Logistics Since the 1998–2008 Reforms

Despite the reforms, the CCP has continued to face challenges in its efforts to deploy and enhance the PLA's logistics support systems. In addition to the enduring challenges that have been discussed previously, such as limited combat experience and the recognition of significant logistics challenges that emerged during that combat experience, three other areas stand out. First, China's military services continued to lack a robust capability to quickly respond and flow necessary supplies throughout the country, such as in the aftermath of natural disasters. Second, corruption has lingered as a major issue that undermines progress in PLA logistics reforms. Third, the PLA has also struggled to provide the necessary training for its personnel.

Maintenance and Other Logistics-Related Constraints on Operations

Responses to natural disasters and the coronavirus disease 2019 (COVID-19) pandemic have highlighted that the PLA services lack logistics capabilities and the capacity to quickly flow materiel. For example, following the 2008 earthquake in Sichuan, PLA command centers lacked the communications capabilities, transport aircraft, and heavy-lift helicopters that were necessary to respond most effectively.[47] Logistics support during the early months of the COVID-19 pandemic revealed some benefits of the JLSF reforms and potential challenges for wartime operations.[48]

Previous RAND research has found that, for day-to-day operations, insufficient amounts of jet fuel and parts, in addition to maintenance times, constrained the PLAAF's annual flight quotas and operations.[49] A lack of heavy

---

[47] Cliff, 2015, p. 150.

[48] Joel Wuthnow, "Responding to the Epidemic in Wuhan: Insights into Chinese Military Logistics," *China Brief*, Vol. 20, No. 7, April 13, 2020.

[49] Edmund J. Burke, Astrid Stuth Cevallos, Mark R. Cozad, and Timothy R. Heath, *Assessing the Training and Operational Proficiency of China's Aerospace Forces: Selections from the Inaugural Conference of the China Aerospace Studies Institute*, RAND Corporation, CF-340-AF, 2016, p. 84.

transports has also presented challenges for the PLAAF's ambitions, including those ambitions for operations abroad.[50] Although the PLAAF has been rapidly increasingly its strategic airlift fleet of Y-20 aircraft, it has only about 50 to date.[51] Furthermore—and unlike the PLA's past experience in combat during the Korean War and other wars—evolving technological advances in military equipment could outpace the PLA's repair and maintenance capabilities.[52] In Chapter 4, we explore how this dynamic might be evolving.

Corruption and Divestiture

Corruption, especially within the PLA logistics system, has been a significant issue that challenges China's military modernization objectives. Although the 1998 reforms began alongside guidance from former Chinese leader Jiang Zemin that the PLA would divest from nonmilitary businesses, corruption purges continued during the Xi era. As part of Xi's anti-corruption campaign, several GLD officers were placed under investigation for corruption, with one essay noting that "among the nearly 50 'tigers'" caught up in investigations, many came from the GLD or were "in charge of approval of real-estate projects."[53] One prominent example of an officer placed under investigation, Gu Junshan, a former deputy director of the GLD, was arrested on bribery charges for selling military positions.[54]

Personnel Issues

Research on PLA logistics has also drawn attention to challenges that stem from a lack of investment in professional opportunities for PLA logistics personnel. Analysts have noted an insufficient number of PLA personnel who work on logistics issues, in particular. For example, Puska notes that the cultivation of logistics personnel has "been hampered by a bias for inde-

---

[50] Garafola and Heath, 2017, p. 30.

[51] International Institute for Strategic Studies, *The Military Balance 2023*, 2023, p. 243.

[52] Cliff, 2015, p. 161.

[53] James Mulvenon, "PLA Divestiture 2.0: We Mean It This Time," *China Leadership Monitor*, No. 50, July 9, 2016, p. 4.

[54] "China Investigates Senior Military Logistics Officer," Reuters, October 21, 2015.

pendent and redundant support provided by each service, which would narrow the logisticians' knowledge and experience."[55] We examine this topic in greater detail in Chapter 4.

The People's Liberation Army's Organizational Approach to Logistics and Maintenance

The greatest impact on logistics of 2015–2016 PLA-wide reforms appears to have been the further development of the PLA's joint logistics apparatus through the creation of the JLSF, which focuses on providing joint logistics services to reduce cross-service redundancies and realize efficiencies, particularly for combat service support.[56] Interestingly, however, most PLA maintenance tasks prior to and following the establishment of the JLSF continue to be anchored within service-specific silos. Based on available information, the JLSF's increasing remit includes responsibilities covering common-use or general-purpose military materiel and food, fuel, medicine, ammunition, and fabricated structures, including personnel shelters. However, that remit does not appear to cover maintenance and servicing of service-specific platforms, such as aircraft and armor.[57] Figure 2.3 identifies the possible JLSF organizational relationship with TCs and PLA service-branch leads within those five different TCs.

This organizational arrangement indicates that Theater Service (Branch) Support Departments (战区军种保障部) are the primary locus for mainte-

---

[55] Puska, 2010, pp. 568–569. Also see Cliff, 2015, p. 149. Limited expertise of personnel has been raised as an issue within other areas of the PLA. See Wuthnow et al., 2021, pp. 11–12. In an expeditionary context, see Garafola et al., 2022.

[56] Luce and Richter, 2019; Office of the Secretary of Defense, *Annual Report to Congress: Military and Security Developments Involving the People's Republic of China*, Department of Defense, May 2022, p. 74; Wuthnow, 2021.

[57] For examples of the work that the JLSF does, as discussed by the PLA media, regarding food, facility fabrication, and medicine, respectively, see "Information Technology Helps Improve Military Food Supply," *China Military Online*, March 15, 2021; Liu Yibo and Chen Pengfei, "Inflatable Camping Tents Distributed to Plateau Troops," *China Military Online*, May 7, 2021; and "Chinese Peacekeeping Medical Contingent to Mali Passes Pre-Mission Test," *China Military Online*, July 5, 2021.

nance activities, such as equipment overhauls and routine engine servicing. For a more thorough listing of PLA maintenance activity types and details, see Table 2.2.

As opposed to earlier periods, the PLA's five TCs do not exercise direct authority over the five geographically aligned JLSF support centers. The TCs have a coordinating relationship with their logistics leads; in turn, the logistics leads provide mobility, fuel, food, medicine, some facility fabrication, and possibly other general-purpose (通用) items, such as certain types of ammunition, to the TCs. PLA reporting in the postmilitary reform era does not typically associate maintenance requirements with JLSF responsibilities.[58] Instead, a preponderance of PLA and state media indicate that maintenance personnel and tasks remain the remit of the TC service branches and that their duties and roles are to perform service-specific (专用) support. See Table 2.2 for more information about these roles.

PLA reporting from China's Ministry of National Defense and *PLA Daily* in 2020 and 2022 typically focused on service-specific efforts and achievements and indicated that, within a designated TC, service-specific maintenance units performed maintenance tasks. These tasks were performed at echelons as high as PLAA Theater Support Departments (战区陆军保障部), which are directly assigned to the TCs,[59] and as low as PLAA battalions in which "grassroots maintenance" (基层维修) is performed by individual sol-

---

[58] In a 2018 layout of the JLSF's remit, maintenance of general-purpose (通用) PLAA equipment was mentioned, but we did not see this reflected in subsequent discussion of PLAA-relevant maintenance activities. See Liu Wenkai [刘文开], Wang Hui [王晖], Ma Yali [马雅丽], He Peng [何鹏], and Chen Shaoshan [陈绍山], "Overall Conception of Theater Army Equipment Support Classification" [战区陆军装备保障体系建设总体构想], *Journal of the Academy of Armored Force Engineering* [装甲兵工程学院学报], Vol. 32, No. 6, 2018, p. 4. Arostegui and Sessions, in their excellent analysis of PLAA logistics, find coordination between Army units and JLSF units on supply and transportation, as well as some hospital linkages for medical treatment (Joshua Arostegui and James R. Sessions, "PLA Army Logistics," *PLA Logistics and Sustainment: PLA Conference 2022*, U.S. Army War College Press, 2023).

[59] See Zhang Hongbo [张洪波] and Lei Ziyuan [雷梓园], "Breaking Through the 'Ceiling' of Technology, Feng Jianzhong Jumped from an Ordinary Soldier to a Maintenance Expert" [冲开技术的 "天花板",冯建中从普通一兵跃升为维修专家], *PLA Daily*, June 27, 2019.

FIGURE 2.3

Possible Postreform People's Liberation Army Logistics and Maintenance Organizational Structure

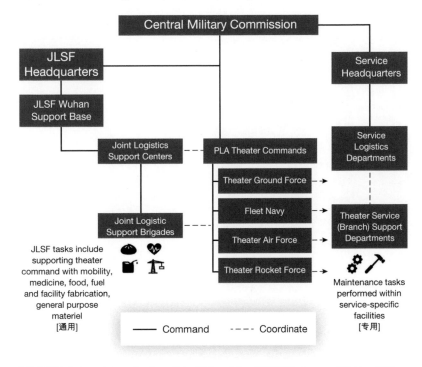

SOURCES: Features information from China Military Online [中国军网].; Luce and Richter, 2019; McCauley, 2018; Joel Wuthnow, "Joint Logistic Force Support to Theater Commands," *PLA Logistics and Sustainment: PLA Conference 2022*, U.S. Army War College Press, 2023, p. 20.

diers at workbenches.[60] Support brigades, battalions, and companies, which typically provide direct support one to two echelons above their unit, were commonly mentioned unit configurations in the context of maintenance

---

[60] PLA sources identify *grassroots maintenance* as maintenance that occurs at or below regiment levels. See Academy of Military Science All-Military Military Terminology Committee [全军军事术语管理委员会，军事科学院], *Chinese People's Liberation Army Military Terminology* [中国人民解放军军语], Academy of Military Science Publishing House [军事科学出版社], 2011, p. 550. For an example of reporting on grassroots maintenance, see Xu Xu [徐徐], Sun Qilong [孙启龙], and Xing Zhe [邢哲], "From One Person to a Team, Explore the 'Correct Way to Open' Equipment Maintenance Support" [从一个人到一个团队，探寻装备维修保障的"正确打开方式"], *PLA Daily*, May 28, 2021.

TABLE 2.2

Reported People's Liberation Army Maintenance Activities, by Type and Echelon

Military Activity Type	Maintenance Activity	Echelon(s) Mentioned
Exercises	Support for and rapid response to equipment failure: In concert with operational units, maintenance teams conduct operations intentionally or incidentally when encountering maintenance issues. Example: Armor brigades attempt repair or replace techniques to restore mobility in timed events.[a]	Company-level maintenance team in support of brigade-level exercise, mobile detachment (People's Armed Police [PAP]), Army Aviation brigade, armor brigade, battalion-level service support to a brigade-level exercise
	Equipment field tests: For persistent equipment problems, or following equipment modifications, maintenance personnel might participate in equipment trials. Example: Operators alternate speeds for a particular power train while maintenance personnel observe equipment effects.[b]	Company-level maintenance team in support of brigade-level exercise equipment test and evaluation unit, armor brigade
	International exercises and competitions: Maintenance personnel participate in competitive exercises held with regional allies for operations and maintenance.[c]	Maintenance personnel organic to an Army Aviation brigade, artillery company, armor brigade
In-garrison skill development	Maintenance manual issuance: Disseminate and review equipment manuals and provide professional development time for junior personnel to familiarize themselves with maintenance methods.[d]	Maintenance personnel organic to a service support battalion
	Revising and disseminating maintenance best practices: As units discover methods to improve equipment maintenance practice, they might facilitate best practice sharing sessions or generate ad hoc maintenance newsletters or flyers.[e] They might also participate in self-study.[f]	Maintenance personnel organic to mobile detachment (PAP), armor brigade

Table 2.2—Continued

Military Activity Type	Maintenance Activity	Echelon(s) Mentioned
	Professional certification: Maintenance personnel study, apply for, and undergo examination for military and commercial maintenance qualifications. Example: Personnel might obtain a riveter's or welder's license.[b]	Maintenance personnel organic to mobile detachment (PAP), service support battalion
	Instructor or expert maintenance lecture: Maintenance personnel receive talks from commercial or military experts in the field of maintenance[h] or receive on-the-job training from senior noncommissioned officers (NCOs).[i]	Maintenance personnel organic to a brigade, Contractor Maintenance Support Unit (PLAAF)
	Operator-maintenance personnel shadowing: Maintenance personnel spend some amount of time following operators in their normal duties, or operators are briefly embedded with maintenance units, to better understand demands placed on equipment and exchange expertise.[h]	Maintenance personnel organic to a brigade
	Post mortem equipment review: Maintenance personnel review video footage of accidents, casualties, and damage caused by incomplete maintenance practices.[g]	Maintenance personnel organic to an Army Aviation brigade
Operational support	Postmission maintenance or service: Prescribed equipment maintenance conducted after aircraft, tanks, etc. operate for a set period (hours) or operate across a set distance (kilometers).[j]	Maintenance personnel organic to a brigade (PLAAF), armor brigade

Table 2.2—Continued

Military Activity Type	Maintenance Activity	Echelon(s) Mentioned
	Emergency maintenance: Units or equipment in extremis receive emergency maintenance support. Example: Damaged but still operable aircraft receive emergency maintenance support.[g]	Maintenance personnel organic to an Army Aviation brigade
	Routine diagnostic checks: Preventive maintenance of equipment conducted regularly, potentially with the assistance of electronic testing equipment.[k] Example: Check aircraft wings for indications of stress cracks.[l]	Maintenance personnel organic to brigade (PLAAF), armor brigade
	Equipment overhaul: Replacement or more-thorough maintenance associated with lifetime extension of equipment.[d] Overhauls might be seasonally determined when unit downtime is anticipated.[m] Example: Replace gearbox, swap engine.	Maintenance personnel organic to mobile detachment (PAP), organic to a brigade
	Equipment acceptance checks: Maintenance personnel review equipment for flaws and check manual-defined standards, potentially executing software or upgrade improvements and debugging issues identified between issuance and unit acceptance.[b]	Maintenance personnel organic to brigade, equipment test and evaluation unit
	Equipment modification: Ambitious or especially talented PLA personnel might receive dispensation (or take unlicensed initiative) to program or tinker with existing issues to overcome issues not solved by original equipment manufacturers. Example: Adapt equipment for inclement weather or reprogram equipment to meet power consumption concerns.[g]	Maintenance personnel organic to a brigade

Table 2.2—Continued

Military Activity Type	Maintenance Activity	Echelon(s) Mentioned
	Troubleshooting equipment failures: For equipment failures without a diagnosed part failure, maintenance personnel conduct a differential diagnosis to determine the cause of inoperability.[n]	Equipment test and evaluation unit, maintenance personnel organic to company, combined arms battalion, armor brigade
	Equipment inventory and database entry: Checking and inventorying equipment and reviewing the equipment faults history, potentially conducted in tandem with troubleshooting equipment failures.[o] Example: specification flowchart.[p]	A repair company providing maintenance support to a battalion, a brigade, a service support battalion providing maintenance support to a brigade

NOTE: The sources from 2018–2022 cited in this table mention "maintenance" (维修), "repair" (修理), and aircraft "service" (服务). Echelons mentioned are assumed to be PLA ground force units unless otherwise mentioned in parentheses, and units were typically referenced as a pairing of two echelons within the PLA source describing the maintenance activity—the higher echelon (e.g., army group, brigade, battalion) receiving support and the lower echelon (e.g., battalion, company) providing direct support.

[a] Chen Dianhong [陈典宏] and Huang Yuanli [黄远利], "Second Sergeant Major Gan Zhanyong: Use a Pair of 'Iron Sand Palms' to Become a Tank 'Maintenance Master'" [二级军士长甘战永: 用一双"铁砂掌"练成坦克"维修大拿"], PLA Daily, August 17, 2019.

[b] Liu Hanbao [刘汉宝], "Approaching 4 Soldiers 'Craftsmen' in Equipment Maintenance Posts: Dedication and Perseverance in the Rear Area of the Battlefield" [走近4名装备维修岗位的士兵"工匠": 战场后方的执着与坚守], PLA Daily, April 12, 2022.

[c] China Global Television Network, "Now Live: 'Ordnance Master' Maintenance Platoon Relay Race" [正在直播: "军械能手"维修排接力赛], webpage, August 14, 2019.

[d] Che Dongwei [车东伟] and Zhang Tong [张童], "A Brigade of the 72nd Group Army Focuses on Improving the Overall Level of the Equipment Maintenance Team" [第72集团军某旅注重提升装备维修队伍整体水平], PLA Daily, August 17, 2018.

[e] Li Tuanbiao [李团标], Xing Zhe [邢哲], and Lei Zhaoqiang [雷兆强], "An NCO and His 400 Plus Issues of Maintenance Tabloids" [一名士官和他的400多期维修小报], PLA Daily, December 23, 2019.

[f] Zhang and Lei, 2019.

Table 2.2—Continued

[g] Tang Wenyuan [汤文元] and Wu Shike [吴世科], "Li Xiangnan: 20 Years Maintaining Warhawks over a Thousand Times with Zero Mistakes" [李向楠：20年维修战鹰千余次零失误], *PLA Daily*, January 17, 2019.

[h] Wang Yi [王轶] and Cheng Jun [程俊], "A Brigade of the 82nd Group Army Innovated the Optimal Support Model Modular Grouping Helps Infantry Fighting Vehicles (IFV) Gallop" [第八十二集团军某旅创新集优保障模式 模块编组助力战车驰骋], *PLA Daily*, April 2, 2022.

[i] Xu, Sun, and Xing, 2021.

[j] Zhang Yuqing [张玉清], Zhang Lei [张雷], Feng Bin [冯斌], "Record of Bai Wenguo, a Mechanical Technician in the Second Maintenance Squadron of the Air Force's August 1st Acrobatic Team" [擦亮名片扬国威—记空军八一飞行表演队机务二中队机械技师白文国], Xinhua, June 5, 2019.

[k] "Li Bei: Use 'X-Ray Eyes' to Detect the War Eagles' Flaws" [李蓓：用"透视眼"为战鹰探伤], *PLA Daily*, October 25, 2018.

[l] Li Weixin [李伟欣], Wei Yumeng [卫雨檬], Shu Xiquan [殳细泉], and Wu Lihua [吴李华], "'This Is My Fighter Jet!' Salute! Air Force Mechanic" ["这是我的战机!" 致敬！空军机务兵], *PLA Daily* [解放军报], January 7, 2021.

[m] Lei Zhaoqiang [雷兆强], Wen Suyi [闻苏轶], and Li Peijin [李沛锦], "How to Face the New Challenges Brought by the New Position? This 'Ace Maintenance Worker' Gives the Answer" [如何面对新岗位带来的新挑战?这名"王牌维修工"给出答案], *PLA Daily*, April 9, 2021.

[n] Yin Weihua [尹威华] and Zhou Huanquan [周焕权], "Expert Plateau Soldiers Have Maintenance 'Knacks' for Emergency Repair of Vehicles, Becoming a 'Reassuring Pill' for Fellow Soldiers" [高原兵专家维修有"诀抢修车成了战友的"定心丸"], Ministry of National Defense Network, August 28, 2018.

[o] Guo Kexin [郭克鑫] and Hai Yang [海洋], "A Brigade of the 79th Group Army Established an Equipment Maintenance Information Database to Facilitate Precision Support" [第79集团军某旅建立装备维修信息数据库助力精准保障], *PLA Daily*, May 23, 2021.

[p] Chen Dianhong [陈典宏], Feng Dengya [冯邓亚], and Zhou Yupeng [周宇鹏], "The Maintenance Center Has Become a 'Smart Factory,' and a Brigade of the 75th Army Group Strives to Improve the Level of Equipment Management Informatization" [维修中心变"智能工厂"，第75集团军某旅着力提升装备管理信息化水平], *PLA Daily*, [解放军报], March 2, 2022.

responsibilities in these publications. These same sources did not colocate JLSF support activities with the TC Service Support Departments, suggesting that in practice, the JLSF's remit does not yet cover maintenance duties that touch individual PLA units' service-specific equipment.

Current People's Liberation Army Maintenance Activities

We next sought to understand the types of activities and operations frontline maintenance units conduct as part of the day-to-day operations. Table 2.2 summarizes maintenance activities described in recent (2018–2022) PLA reporting. Identified PLA maintenance activities disproportionally represent the PLA ground forces, occasionally mention PLAAF, PLAN, PAP, and commercial contractor maintenance efforts and entirely omit any PLARF maintenance activities.[61]

Table 2.2 samples self-reported PLA activities and does not reflect a frequency count of activities, although troubleshooting (排除故障) equipment failures was one of the most common mission support vignettes featured across TCs. Revising and disseminating maintenance best practices were also common PLA maintenance vignettes highlighted in in-garrison training.

To summarize some common themes in each category, we found reporting on PLAA and PLAAF units that practiced maintenance-specific training and on-site repairs, such as rapid repair in a field exercise for the PLAA and examining maintenance challenges during extreme cold weather for the PLAAF.[62] The variety of in-garrison skill training included professional military education and practical opportunities. In the PLAA, maintenance units helped fill technical gaps in manuals and repair guides and

---

[61] The seeming absence of PLARF maintenance activities from open-source media and official Chinese sources is unsurprising; the PLARF's remit covers sensitive aspects of China's nuclear weapons arsenal and delivery systems. The emphasis on PLA ground forces might be a reporting bias of PLA sources and might reflect some level of sensitivity in reporting about air and naval platforms.

[62] Yang Pan [杨盼], "Cold Training: What to Do If the War Eagle Is Cold? The Air Force Pilots Who Participated in the Cold Training Will Tell You" [寒训 | 战鹰冷了怎么办?参加寒训的空军机务兵告诉你], *China Military Online* [中国军网], January 4, 2019.

conducted exchanges with local academic institutions. PLAAF activities included sending maintenance schoolhouse instructors to combat units for internships or rotations and creating a unit that supported the compilation of a textbook on a "new bomber's" maintenance by providing maintenance data and recommendations to the textbook authors.[63] The PLAA and PLAAF units both developed tools to shorten maintenance times.[64] In terms of operational support, articles on the PLAA emphasized building the skills of junior technicians, and articles on the PLAAF focused on the transition away from maintenance "down days" to a "precision" inspection-based maintenance model.[65] In Chapter 4, we describe these dynamics in greater detail, as well as their implications for thinking through how the PLA approaches maintenance.

Beyond Day-to-Day Operations: Maintenance's Links to Other Organizations

PLA maintenance activities could also increasingly link up with efforts to modernize China's national defense mobilization system.[66] China's national defense mobilization (国防动员) strategy, outlined in a 2010 People's Republic of China (PRC) law, is focused on converting latent national means of materiel production and logistics capacity to support wartime needs.[67] National defense mobilization is managed by the National Defense Mobi-

---

[63] On the "new bomber" textbook, see Xia and Yang, 2021.

[64] Burke et al., 2016, p. 76; Chen, Feng, and Zhou, 2022; Guo and Hai 2021.

[65] Kenneth W. Allen and Cristina L. Garafola, *70 Years of the PLA Air Force*, China Aerospace Studies Institute, April 12, 2021, p. 270; Burke et al., 2016, p. 77; Zhang Zhaoyang [张照洋] and Huang Kainan [黄凯楠], "A Brigade of the 72nd Group Army: Practicing New Equipment and Strengthening Old Equipment" [第72集团军某旅：练熟新装备 练强老装备], *PLA Daily*, April 23, 2022.

[66] For discussion on the links between logistics and national defense mobilization that are relevant in a Taiwan scenario, see Chieh and Yang, 2021.

[67] Officially, the "National Defense Mobilization Law of the PRC" [中华人民共和国国防动员法] adopted by the Thirteenth Meeting of the Standing Committee of the Eleventh National People's Congress on February 26, 2010 ("National Defense Mobilization Law of the PRC (Full Text)" [中华人民共和国国防动员法 (全文)], Xinhua [新华网], February 26, 2010).

lization Commission (国防动员委员会), which is led by the Central Military Commission (CMC) and the State Council of the PRC in an effort to create national unity across civilian and military entities during a time of war. China's mobilization system consists of seven pillars, or seven lines, of mobilization efforts: science and technology, armed forces, people's air defense, political, information, economic, and equipment.[68] Equipment is conceivably the area in which PLA maintenance might have overlap or synergy with existing mobilization efforts to marshal civilian resources to support wartime needs.[69]

Mobilization committees and a mobilization chain of command extend from the CMC-State Council down to the TCs in successively lower echelons, with offices and committees reaching the county level. As of 2019, one PLA source reported that mobilization committees were fully established at the national and provincial levels and had not yet been established at the TCs.[70] Mobilization staff who hold offices within the TCs are tasked with conducting an annual standardized survey of their regions' contributions to the seven mobilization categories to account for railheads, bridges, motor pools, and available hospital beds, among many other categories. [71]

These sources do not make clear whether more-complicated pieces of equipment and maintenance support are covered in these surveys. In a broad sense, PLA accounting for cots, pallets, and tents includes certain types of equipment but excludes technologically significant equipment that has maintenance implications across the force. Some sources that have prescribed or critiqued the use of mobilization information systems to collect and monitor mobilization resources do extend to categories of equipment

---

[68] Dean Cheng, "Converting the Potential to the Actual: Chinese Mobilization Policies and Planning," in Andrew Scobell, Arthur S. Ding, Phillip C. Saunders, and Scott W. Harold, eds., *The People's Liberation Army and Contingency Planning in China*, National Defense University Press, 2015.

[69] On this point, see Wuthnow, 2023, p. 21.

[70] Chen Wengang [陈文刚] and Hou Biao [侯彪], "A Preliminary Inquiry into the Building of National Defense Mobilization Command Authorities in the Provincial Realm" [省域国防动员指挥机构建设初探], *National Defense* [国防], No. 7, 2019.

[71] Song Naihong [宋乃宏], "Some Thoughts on Doing Well in the Investigative and Statistical Work on National Defense Mobilization Potential in the New Era" [新时代做好国防动员潜力统计调查工作的几点思考], *National Defense* [国防], No. 8, 2019, p. 3.

support maintenance (装备维修保障). One source states that "equipment mobilization includes civil equipment requisition [and] equipment maintenance and repair," perhaps suggesting that mobilization efforts stipulate maintenance and repair capabilities at the programmatic level.[72] As we discussed in the "Organization" section, in the PLA literature, there are few ties between these types of activities and the articulation of the JLSF's role in national mobilization efforts, even though the similarity in kinds of materiel support, fuel, transportation, medicine, and other mobilization efforts would seem to have some complementary function with JLSF efforts in a wartime contingency.

Conclusion

The PLA's drive toward logistics reforms has been motivated by historical challenges, such as logistics failures during the Chinese military's participation in previous conflicts, observations of advances in Western militaries, and desires to realize greater efficiencies within the logistics system. However, even following the 1998–2008 logistics reforms, operationalizing logistics capability rapidly throughout the country was challenged by the need to curb corruption and provide sufficient personnel training for logistics. The 2015–2016 PLA-wide reforms introduced broad changes to PLA logistics in the form of the JLSF. However, a review of the position of maintenance organizations and units within the broader organizational structure and descriptions of unit-level maintenance activities indicate that most maintenance activities remain service-specific functions that are separate from the JLSF activities that are carried out by TC Service Support Departments.[73] Our survey of PLAA and PLAAF maintenance activities identified several activities to facilitate exercises, promote in-garrison skills development, and

---

[72] Wang Jizhen [王纪震] and Ding Guandong [丁冠东], "Equipment Mobilization Information System Construction" [装备动员信息系统构建], *Command Information System and Technology* [指挥信息系统与技术], Vol. 9, No. 3, 2018.

[73] Continued reforms to the JLSF could potentially alter this arrangement. For PLA analyst discussion of other potential reform options for PLA joint logistics, see Wuthnow, 2023.

provide operational support and set the stage for our exploration of what PLA maintenance challenges and successes imply about PLA maintenance organizational culture in Chapter 4.

Overall, one thread that weaves through both the historical drivers and the PLA's resulting adaptations to its logistics and maintenance structures is a self-assessment of the need to compensate for existing weaknesses or inferiorities compared with the forces and capabilities of potential opponents. However, as M. Taylor Fravel wrote in 2019, "changes in China's material power in the past decade raise serious questions about the meaning of active defense," which is China's overarching military strategy and is based on long-standing assessments about the PLA's inferiority compared with opponents, "because China will no longer be in a position of material inferiority."[74] Specifically, it is unclear how the PLA's approach to compensation might be tested in relation to maintenance as China continues to modernize its military hardware; some PLA weapon systems and platforms are currently world class or approaching world-leading status. In Chapter 3, we lay out our framework for understanding maintenance success factors. In Chapter 4, we identify areas of progress and continued deficiencies that are shaped by the PLA's approach to maintenance management.

---

[74] M. Taylor Fravel, *Active Defense: China's Military Strategy Since 1949*, Princeton University Press, 2019, p. 63.

A Framework for Exploring Maintenance Success Factors

In this chapter, we develop a framework for exploring the links between broader attributes of organizational culture and successful maintenance practices. The chapter presents logistics success factors that can inform the analysis of trends in maintenance approaches, along with examples of both successes and failures from historical military operations.[1] Our initial review resulted in 11 factors, but we refined those factors to six as we discovered which aspects were most relevant for analyzing PLA maintenance. This process included focusing our efforts where we saw strong themes emerging but excluding those factors for which we noticed little mention of in PLA sources. The exclusion of those five factors does not mean that they are entirely irrelevant but that they appear to be much less relevant than the six factors that we present in this chapter. The remaining five factors are detailed further in Appendix A.

Defining a Logistics Culture

An organizational culture spans the entirety of an organization, explains why organizations operate the way they do, and ultimately affects organiza-

---

[1] Because of the limited discussion within the literature related to organizational culture and maintenance, we broadened our search to look at what the literature contained related to organizational culture and logistics. We then evaluated which aspects of a logistics organizational culture would be most relevant and applicable to maintenance.

tional effectiveness.[2] Organizations—and their associated culture—can be examined at different levels to explore effectiveness. For example, we can broadly consider the culture of military organizations; look more specifically at a given country's military culture; or, focusing further, examine the culture of a particular functional or operational unit (e.g., the U.S. Army's maintenance organization). Because of the limited channels that are available to gain a holistic picture of the PLA's logistical capabilities, we explored the concept of a logistics culture as a stepping stone to understanding the PLA's logistics objectives and its potential strengths and weaknesses. Specifically, we combined aspects of organizational theory, military studies, and historic case analyses to develop a framework for assessing the potential strengths and weaknesses of a military logistics organization.

The framework includes 11 factors associated with successful logistics cultures, which are summarized in Table 3.1. Because we focused our study on the PLA's maintenance organizational approach and effectiveness, we further identified a subset of six of these logistics success factors by either their relevance to the maintenance subfunction of logistics or their prevalence in the Chinese literature. We elaborate on these six factors in the following sections and explore the remaining five factors in Appendix A. We found that the available, relevant research specific to military logistics management, and maintenance more specifically, was limited. When previous research on logistics or the maintenance subfunction was available, we incorporated it in our development of each of the 11 factors, but we also supplemented our analysis with studies on civilian organizations (such as the commercial airline industry) and other military functional areas (such as information technology). We acknowledge that the factors below are not exhaustive; other factors likely contribute to the success or failure of a military logistics organization. However, taken together, these factors provide an important starting point for assessing a logistics organization's capabilities.

---

[2] Nathan Voss and James Ryseff, *Comparing the Organizational Cultures of the Department of Defense and Silicon Valley*, RAND Corporation, RR-A1498-2, 2022.

TABLE 3.1

Factors Present in Successful Logistics Organizational Cultures

Factor in Success	Description of Factor and Results	Examples of Operational or Strategic Success or Failure
Culture of learning	• Continual analysis of logistics procedures • Investment in new technologies or processes during peacetime • Application of lessons learned	• U.S. Army's Center for Army Lessons Learned, Haiti, 1994 • Chinese People's Volunteer Force • Korean War, 1950–1953
Agility leveraging the civilian sector	• Ability to leverage civilian infrastructure • Knowledge-sharing between civilian and military organizations • Interoperability between civilian and military sectors	• The U.S. Department of Defense's (DoD's) application of the Defense Production Act for Mine-Resistant Ambush-Protected vehicles, 2000s • Vietnam War, Viet Cong and North Vietnamese Army forces' integration of civilians into a military logistics system, 1955–1975
Routinization and prioritization of maintenance functions	• Maintenance of equipment on a regular basis • Allocation of sufficient time to perform essential equipment maintenance • Documentation and practice of maintenance triage • Processes to ensure that inspections and maintenance are performed to standard	• Russian invasion of Ukraine, 2022 • Flight operations on U.S. Navy carriers, 1980s

Table 3.1—Continued

Factor in Success	Description of Factor and Results	Examples of Operational or Strategic Success or Failure
Latitude for independent decisionmaking by logisticians at the unit level	• Reliance on logisticians to exercise independence and judgment in carrying out their duties rather than following strict rules • System of maintenance management that reflects the needs of performing organizations as opposed to the objectives of the controlling agencies	• Flight operations on U.S. Navy carriers, 1980s
Cradle-to-grave view of sustainment	• Life-cycle management (LCM) approach to acquisition that incorporates logistics from the initial design of a system • Level of collaboration between acquisition and sustainment branches	• DoD acquisition reforms, 1980s–1990s • Defense LCM in the PLA's General Armament Department (GAD), 2010s
Ability to recruit and retain quality personnel	• Provision of benefits and incentives that affect morale and retention to military personnel, such as adequate housing, food, and wages • Resolution of issues related to unequal treatment of different groups • Prioritization of training and education in a given field or specialty, such as training in specific maintenance tasks	• U.S. transition to an all-volunteer force, 1970s

Table 3.1—Continued

Factor in Success	Description of Factor and Results	Examples of Operational or Strategic Success or Failure
Integration of logistics personnel into operational units[a]	• Colocation of logistics personnel with operational units • Use of military exercises that include both operations and support elements working together in realistic scenarios • Integration and coordination of logistics plans with strategic plans	• United States in the Pacific War, 1941–1945 • World War I, Mobile Ordnance Repair Shops, 1917–1918
Unified chain of command[a]	• Unified command structure with a clear chain of command • Clear lines of communication • Span of control	• U.S. military information operations, 2010s
Appropriate balance between competing priorities in a constrained environment[a]	• Appropriately balanced cost, quality, and timeliness factors related to maintenance and supply • No prioritization of cost, quality, or timeliness factors at the expense of each other	• Flight operations on U.S. Navy carriers, 1980s
Low prevalence of corruption[a]	• Lack of corruption that diverts scarce resources and has negative effects on operational performance	• Russian invasion of Ukraine, 2022
Balanced emphasis between support and combat elements[a]	• Sufficient support to combat soldier ratio • Protection provided to support units	• Russian invasion of Ukraine, 2022 • United States in the Pacific War, 1941–1945

[a] This factor is discussed further in Appendix A.

43

Culture of Learning

Organizational learning refers to a change in an organization's knowledge in the form of "cognitions, routines, and behaviors" that occurs because of experience.[3] More specifically, organizational learning consists of "creating, retaining, and transferring knowledge."[4] An organization's learning process can provide important insight into its effectiveness and how it sees itself in relation to other organizations. For example, Bierly and Chakrabarti describe how commercial organizations learn from either their own experience or the experience of other organizations and suggest that successful firms place equal emphasis on both types of learning.[5] The evolutionary stage of an organization might also affect which of these two types of learning the organization gravitates toward. For example, newer and less-established organizations depend more on other organizations for learning than they do on their own experience.[6] Furthermore, organizations tend to adopt outside programs that are perceived as successful; the rate of adoption of new programs tends to be higher when an organization perceives its own performance to be lacking.[7] Although the literature on organizational learning focuses on civilian firms, some researchers have applied these concepts to a military context. Laksmana explores the concept of military mimicry in Cold War Indonesia and Meiji Japan and concludes that foreign militaries can exhibit varying levels of emulation depending on several factors.[8]

---

[3] Linda Argote, "Organizational Learning Research: Past, Present and Future," *Management Learning*, Vol. 42, No. 4, September 2011, p. 440.

[4] Argote, 2011, p. 440.

[5] Paul Bierly and Alok Chakrabarti, "Generic Knowledge Strategies in the U.S. Pharmaceutical Industry," *Strategic Management Journal*, Vol. 17, No. S2, Winter 1996.

[6] Amalya L. Oliver, "Strategic Alliances and the Learning Life-Cycle of Biotechnology Firms," *Organization Studies,* Vol. 3, May 2001.

[7] Matthew S. Kraatz, "Learning by Association? Interorganizational Networks and Adaptation to Environmental Change," *Academy of Management Journal*, Vol. 41, No. 6, December 1998.

[8] Evan Abelard Laksmana, *Imitation Game: Military Institutions and Westernization in Indonesia and Japan*, dissertation, Syracuse University, 2019.

In the context of military logistics, a culture of learning applies to such practices as the continual analysis of logistics procedures, investment in new technologies or processes during peacetime, and the application of lessons learned. After-action reports are one mechanism that has been found to be successful in institutionalizing continuous learning and improvement in military organizations.[9] Broadly, this process consists of "review[ing] the intent of an operation, analyz[ing] what happened, captur[ing] the lessons learned and their implication for future action, and apply[ing] the lessons quickly back into action."[10] Kane describes how logistics innovations during the interwar period ultimately enabled the U.S. Navy to defeat Japan in the Pacific War.[11] In part, these innovations were a response to the lessons of World War I, including the Allied fleet's failed campaign in the Dardanelles, and consisted of both technical changes (e.g., the adoption of amphibious tractors) and procedural changes (e.g., the way cargo was positioned within a ship). During the Korean War, General Hong Xuezhi of the CPVF demonstrated a culture of learning by establishing a new logistics system based on a network of supply depots positioned along the front lines.[12] Prior to the introduction of this system, food and munitions were delivered to the front line via an inefficient process that involved numerous levels and transactions, from CPVF army headquarters to their divisions, and from the divisions to the regiments. During the first three campaigns of the war, prior to the introduction of the system, the CPVF met between only 25 percent and 40 percent of its troops' food needs. Following the introduction of Hong's system, the CPVF was able to significantly improve its ability to deliver essential supplies, increasing frontline troops' combat effectiveness.

On the question of the PLA's logistics capabilities, the literature raises several questions, the answers to which could point to the potential strengths or weaknesses of the PLA's logistics organization. For example,

---

[9] Lloyd Baird, John C. Henderson, and Stephanie Watts, "Learning from Action: An Analysis of the Center for Army Lessons Learned (CALL)," *Human Resource Management*, Vol. 36, No. 4, December 1997.

[10] Baird, Henderson, and Watts, 1997, p. 387.

[11] Thomas Kane, *Military Logistics and Strategic Performance,* Taylor & Francis, 2001.

[12] Xiaobing Li, *A History of the Modern Chinese Army*, University Press of Kentucky, 2007.

to what extent do PLA logistics organizations balance learning from other organizations with their own experiences? What can this tell us about the PLA's performance and level of organizational maturity? Have continuous learning and improvement processes related to logistics and the maintenance subfunction been institutionalized, or do changes happen in a more decentralized and ad hoc manner? What technologies, if any, is the PLA investing in to improve its logistics processes?

Agility Leveraging the Civilian Sector

Militaries can be enabled—or constrained by—the health of the associated civilian sectors, from which they can draw personnel, infrastructure, services, and technology. Military coordination with the civilian sector can take various forms, including the incorporation of military standards into civilian construction (such as transportation and communication infrastructure; joint active duty and civilian training) and collaboration with civilian scientific research organizations on emerging technologies with military applications.[13] An example of civilian-military coordination from the United States is DoD's use of the Defense Production Act during the initial years of the Iraq and Afghanistan conflicts to prevent a shortage of armor plates for Mine-Resistant Ambush-Protected vehicles.[14] In the context of budgetary and manpower constraints, Ritschel and Ritschel find that the ability to integrate civilian capabilities might have a particularly significant impact on military maintenance activities.[15] Through a statistical analysis of the relationship between maintenance strategies and performance metrics across the U.S. Air Force's aircraft platforms, the researchers

---

[13] Kevin McCauley, written testimony for hearing on China's Military Power Projection and U.S. National Interests, U.S.-China Economic and Security Review Commission, February 20, 2020.

[14] Emma Watkins and Thomas Spoehr, *The Defense Production Act: An Important National Security Tool, But It Requires Work*, The Heritage Foundation, October 15, 2019.

[15] Jonathan D. Ritschel and Tamiko L. Ritschel, "Organic or Contract Support? Investigating Cost and Performance in Aircraft Sustainment," *Journal of Transportation Management*, Vol. 26, No. 2, Fall–Winter 2016.

find that contracted support outperforms organic support in achieving key aircraft availability targets.

Although there are strategic and operational benefits associated with leveraging the civilian sector in pursuit of military objectives, such benefits are not necessarily automatic. For example, Wentz found that coordination between civilian and military units is essential to the successful planning and execution of disaster and emergency responses but that several issues—including "tensions between military needs for classification (secrecy) of data, versus the civilian need for transparency" and "differences in the command-and-control style of military operations versus civilian activities"—can complicate effective civilian-military coordination.[16] Based on a survey of about 3,500 military and civilian personnel from five NATO partner countries, Goldenberg et al. found that "positive military-civilian personnel work culture and relations" require respect, intergroup understanding, familiarity, fairness, and equal treatment.[17] If these conditions are met, the military organization is more likely to benefit from "exposure to new or diverse perspectives of both military and civilians, often due to their different backgrounds, culture, skills, job experience, or history," and "continuity and corporate memory provided by civilian personnel."[18]

In a combat setting, historical case study analyses have underscored the important role that civilian infrastructure plays in winning wars. Kane, for example, describes the Communist forces' ability to integrate civilians into their logistics system, which enabled them to wage a protracted war against the United States and its allies during the Vietnam War. Viet Cong and North Vietnamese Army forces, specifically, were able to establish "a well-organized system of taxation, commerce, forced labor, local transpor-

---

[16] Larry Wentz, *An ICT Primer: Information and Communication Technologies for Civil-Military Coordination in Disaster Relief and Stabilization and Reconstruction*, Center for Technology and National Security Policy, National Defense University, July 2006.

[17] Irina Goldenberg, Manon Andres, Johan Österberg, Sylvia James-Yates, Eva Johansson, and Sean Pearce, "Integrated Defence Workforces: Challenges and Enablers of Military-Civilian Personnel Collaboration," *Journal of Military Studies*, Vol. 8, 2019, p. 28.

[18] Goldenberg et al., 2019.

tation networks," and clandestine depots to sustain Communist fighters."[19] Through their use of civilian resources, the Communists were able to allocate trained soldiers for duty on the battlefield instead of in the logistics system.[20] In China, the integration of the civilian sector into the country's military logistics system has become a key feature of military modernization efforts. For example, in 2009, the PLAN established its first integrated civil-military vessel equipment center, which has since enhanced equipment support to vessels throughout South China.[21] What other pathways might the PLA have to draw expertise, manpower, or other resources from the civilian logistics sector? Are these pathways ones that the United States and other countries would be precluded from accessing because they have different governance or economic systems? What is known about the level and nature of coordination between PLA military personnel and civilians, and what can this tell us about the capabilities of the PLA's maintenance organization?

Routinization and Prioritization of Maintenance Functions

Within an organization, routines serve several key functions. They allow organizations to "coordinate their activities internally" through a shared understanding of divisions of labor and hierarchies.[22] They also enable individuals within an organization to address standard situations without expending unnecessary energy. Finally, routines carry organizational knowledge by embodying the "successful solutions to problems by the organization in the past."[23] Routinization of military maintenance includes such factors as the extent to which equipment is maintained on a regular basis,

---

[19] Kane, 2001.

[20] Kane, 2001.

[21] Puska, 2010.

[22] Cornelius Friesendorf, *How Western Soldiers Fight: Organizational Routines in Multinational Missions*, Cambridge University Press, May 2018, p. 52.

[23] Massimo Paoli and Andrea Prencipe, "Memory of the Organisation and Memories Within the Organisation," *Journal of Management and Governance*, Vol. 7, 2003, p. 150.

the allocation of sufficient time to perform essential equipment maintenance, whether maintenance triage is documented or practiced, and the extent to which there are processes in place to ensure that inspections and maintenance are performed to standard. Underscoring the importance of prioritizing maintenance routines, military analysts have attributed "excessive vehicle breakdowns" as a key logistical challenge that has impeded the ongoing Russian invasion of Ukraine.[24] These analysts note that Russian troops had been conducting military exercises for months before entering Ukraine and did not have time to conduct essential equipment maintenance in advance of the operation.[25]

Research related to safety in civilian organizations provides further insight into the role of routines and standards among maintenance personnel. Patankar, for example, finds that an emphasis on compliance with the standard operating procedures (SOPs) was a key factor contributing to a safety record free of accidents in an aviation organization.[26] Flight crew and maintenance personnel from this organization were more likely to attribute safety to SOPs than were other employees. In a study on the importance of organizational *slack* (defined as "time and human resources that are not constantly subject to measures of short-term efficiency"), Lawson describes how cost constraints at chemical and nuclear plants have resulted in cuts to preventive-maintenance activities, which have the potential for disastrous results.[27] For military organizations, Rochlin, La Porte, and Roberts describe how robust SOPs and procedures aboard aircraft carriers help ensure the reliability of flight operations despite high personnel turnover.[28] Together, these studies speak to the importance of prioritizing and

---

[24] Bonnie Berkowitz and Artur Galocha, "Why the Russian Military Is Bogged Down by Logistics in Ukraine," *Washington Post,* March 30, 2022.

[25] Berkowitz and Galocha, 2022.

[26] Manoj S. Patankar, "A Study of Safety Culture at an Aviation Organization," *International Journal of Applied Aviation Studies,* Vol. 3, No. 1, March 2003.

[27] M. B. Lawson, "In Praise of Slack: Time Is of the Essence," *Academy of Management Executive,* Vol. 15, No. 3, August 2001, p. 125.

[28] Gene I. Rochlin, Todd R. La Porte, and Karlene H. Roberts, "The Self-Designing High-Reliability Organization: Aircraft Carrier Flight Operations at Sea," *Naval War College Review,* Vol. 40, No. 4, Autumn 1987.

standardizing maintenance routines for ensuring safety and operational effectiveness and raise key questions about the PLA's maintenance organization. For example, to what extent are quality control procedures and routine maintenance inspections required, implemented, and documented on a regular basis? Are maintenance groups within the PLA provided with sufficient time and resources needed to effectively carry out maintenance work?

Latitude for Independent Decisionmaking by Logisticians

In his analysis of U.S. naval doctrine—which posits that the most important principle of logistics is providing the "right support, at the right time, in the right place"—Kane reflects on a key principle of military logistics: that logisticians "do their jobs, not by applying precise rules, but by exercising independence and judgement within broad parameters."[29] The literature on the performance of centralized versus decentralized organizations is mixed. However, there is some consensus that a certain level of autonomy facilitates problem-solving and decisionmaking. In a study of aviation line maintenance technicians, Pettersen and Aase posit that flexibility in problem-solving is a prerequisite for the "development of safe work practices."[30] Specifically, they find that technicians' ability to identify and resolve errors in a short amount of time is enabled by a flat organizational structure that allows frontline technicians who have "first hand information about the technical system and [know] about changes in operational conditions" to make flexible decisions.[31] For example, they might directly contact specialists in other departments for assistance, regardless of those individuals' positions in the organization. Similarly, both Pettersen and Aase and Rochlin, La Porte, and Roberts describe the importance of on-the-job train-

---

[29] Kane, 2001, p. 5.

[30] Kenneth A. Pettersen and Karina Aase, "Explaining Safe Work Practices in Aviation Line Maintenance," *Safety Science,* Vol. 46, No. 3, March 2008, p. 513.

[31] Pettersen and Aase, 2008, p. 515.

ing and practical experience in developing a maintainer's skill set.[32] Both studies note that the skills that are needed to address complex maintenance issues, such as interactions between different systems of an aircraft, require knowledge beyond what can be obtained via detailed maintenance manuals or classroom training. In other words, effective maintenance requires autonomy in the form of space for trial-and-error and direct training and retraining.

Autonomy might be particularly salient in what Roberts describes as *high-reliability organizations,* or organizations that use complex and potentially disastrous technologies and tools but "have operated nearly error free for very long periods of time."[33] As Rochlin, La Porte, and Roberts describe in their study of aircraft carrier flight operations, delegation to the lowest level of authority combined with an organizational culture that eschews punishment is a significant factor in explaining the U.S. Navy's high reliability:

> Even the lowest rating on the deck has not only the authority, but also the obligation to suspend flight operations immediately, under the proper circumstances and without first clearing it with superiors. Although his judgment might later be reviewed or even criticized, he will not be penalized for being wrong and will often be publicly congratulated if he is right.[34]

For the PLA, what is known about the incentives in place for maintenance officers and technicians to exercise independent thinking and judgment? What is the health of on-the-job training among PLA maintainers?

Cradle-to-Grave View of Sustainment

About 70 percent of a product's life-cycle costs—the costs of operating and sustaining the product over its lifetime—are cemented during the product's

---

[32] Pettersen and Aase, 2008; Rochlin, La Porte, and Roberts, 1987.

[33] Karlene H. Roberts, "Managing High Reliability Organizations," *California Management Review,* Vol. 32, No. 4, July 1990.

[34] Rochlin, La Porte, and Roberts, 1987, pp. 83–84.

early design stage.[35] For example, DoD plans to acquire nearly 2,500 F-35 aircraft for about $400 billion but projects spending another $1.3 trillion—or 76 percent of the combined acquisition and sustainment costs—to operate and sustain these aircraft.[36] In other words, the cost of sustaining a defense system as calculated in maintenance person-hours, spare parts, upgrades, etc. is defined before the system is even fielded and can make up the majority of a system's total costs. As described by Flanagan, the concept of LCM emerged in DoD in the 1990s to address the "problem of optimized performance decisions made during weapon system development resulting in weapon systems that were difficult and expensive to maintain."[37] Prior to the introduction of LCM concepts, DoD acquisition and logistics were generally siloed, as demonstrated by the separate chains of command for acquisition and sustainment and budgeting structures that further exacerbated divisions between the two communities. In response, DoD designated the weapon system program manager as the entity responsible for LCM, including "acquisition, development, production, training, fielding, sustaining, and disposal of a DoD system," with the expectation that planning for support would start at program inception and supportability requirements would be "balanced with other requirements that impact program cost, schedule, and performance."[38]

Terzi et al. describes similar challenges related to fragmented acquisition and sustainment elements in the commercial sector, which the authors attribute to the growing "quantity, complexity and variety" of products.[39]

---

[35] Benedict Uzochukwu, Silvanus Udoka, Paul Stanfield, and Eui Park, "Design for Sustainment—A Conceptual Framework," *Proceedings of the 2010 Industrial Engineering Research Conference*, January 2010.

[36] U.S. Government Accountability Office, *F-35 Sustainment: DOD Needs to Cut Billions in Estimated Costs to Achieve Affordability*, July 2021.

[37] Michael P. Flanagan, *Life Cycle Management Commands: Wartime Process or Long-Term Solution?* thesis, Army War College, 2007.

[38] Defense Acquisition University, "Product Support Integrator (PSI) and Product Support Provider (PSP)," webpage, undated.

[39] Sergio Terzi, Abdelaziz Bouras, Debashi Dutta, Marco Garetti, and Dimitris Kiritsis, "Product Lifecycle Management—From Its History to Its New Role," *International Journal of Product Lifecycle Management*, Vol. 4, No. 4, November 2010.

Product information that was previously contained within a single entity has become increasingly dispersed as a result.[40] The repair phase, for example, was once performed by "the same entity on well-known items" but is now "dispersed from a spatial and organizational point of view, among many different service units."[41] The solution, the authors conclude, is the increased integration of product information. In the military context, failure to integrate product information can have serious operational impacts. In an analysis of China's defense LCM system, for example, Puska et al. describes how the PLA's GAD—which was responsible for equipping and arming units—had failed to "effectively oversee the development of military weapons and equipment" in the pre-2016 reform era, which led to "persistent problems in quality control and a mismatch between defense production and military user requirements."[42] The authors explain that, in the long run, these shortcomings "slow production and weaken military sustainment, resulting in early obsolescence of weapons and equipment" in the PLA.[43] The question remains as to whether the 2016 reforms have successfully addressed challenges related to a lack of oversight in the development of military weapons and equipment. Do weaknesses in the PLA's life-cycle management system continue to result in maintainability and reliability issues?

Ability to Recruit and Retain Quality Personnel

The recruitment and retention of military personnel has been extensively covered by organizational psychology and related disciplines. Broadly, it is acknowledged that the ability to recruit and retain quality personnel (in terms of personal characteristics or technical skills) is a defining factor in organizational performance. For example, examining the correlation

---

[40] Terzi et al., 2010, p. 362.

[41] Terzi et al., 2010, p. 362.

[42] Susan M. Puska, Aaron Shraberg, Daniel Alderman, and Jana Allen, "A Model for Analysis of China's Defense Life Cycle Management System," *Study of Innovation and Technology in China Policy Briefs*, No. 4, 2014, p. 2.

[43] Puska et al., 2014, p. 2.

between maintainer experience and the readiness level of F/A-18 Super Hornet squadrons, Hatzung and Welborn find a significant, positive relationship between maintainer experience and unit readiness.[44] The authors stress that rapid changes in technology, which outpace the U.S. Navy's ability to train its personnel, exacerbate gaps in experience levels among naval aviation maintainers.[45]

The ability to recruit and retain quality personnel is a product of morale, which in turn reflects workplace culture. Many of these relationships have been long understood to affect military capability. For example, as Carl von Clausewitz observed, "anyone who tries to maintain that wretched food makes no difference to an army . . . is not taking a dispassionate view of the subject."[46] Studies on aviation maintenance and workplace culture have reinforced this observation. For example, based on a survey of 240 maintenance engineers working at the two main helicopter repair bases for the Australian Army, Fogarty finds that both *turnover intention* (whether a respondent intends to keep working in the maintenance industry) and self-reported maintenance errors are driven by morale and individual psychological health. Morale includes such factors as job satisfaction, commitment to the organization, sense of personal responsibility, and individual psychological health includes such factors as exposure to workplace stressors and fatigue. Morale and psychological health are in turn influenced by organizational climate, which can include the level of recognition received for doing good work, the frequency of feedback on work performance, and the quality of training.[47]

In the military context, research on the differences between conscript and volunteer forces has sought to link recruitment strategies to performance on military jobs. These studies were of particular interest to us

---

[44] Scott A. Hatzung and David B. Welborn, *Leveraging Maintainer Experience to Increase Aviation Readiness*, thesis, Naval Postgraduate School, 2020.

[45] Hatzung and Welborn, 2020.

[46] Carl von Clausewitz, *On War*, trans. and ed. by Michael Howard and Peter Paret, Princeton University Press, 1976, as cited in Kane, 2001.

[47] Gerard Fogarty, "The Role of Organizational and Individual Differences Variables in Aircraft Maintenance Performance," *International Journal of Applied Aviation Studies*, Vol. 4, No. 3, March 2004.

because military service is technically obligatory for Chinese nationals, although many slots appear to have been filled with voluntary applications in recent years.[48] Of those voluntary applicants, the PLA has likewise sought to recruit a greater proportion of college graduates and skilled personnel through various incentive programs.[49] In the case of the United States, research indicates that an all-volunteer force has managed to recruit individuals with a high school diploma or with Armed Forces Qualification Test scores at or above the median in the youth population at a higher rate than during the Vietnam War through a draft.[50] Both high school completion and higher Armed Forces Qualification Test scores have been found to be predictors of better performance in military jobs.[51] Adding to this, Burk notes that volunteer militaries have more flexibility in letting go of low performers compared with conscript militaries, which encourage stricter discipline.[52] He also notes that freedom to select who should join and stay in the military "narrows the range and diversity of values represented in the force" and therefore contributes to a more uniform force. Applying this research to the PLA, how successful has the organization been historically in recruiting and retaining qualified maintainers? What barriers, if any, are there to maintaining a skilled cadre of technicians? What is the potential impact of increasingly complex military technology on the recruitment and retention of PLA maintainers?

---

[48] On this dynamic and for more information on the conscription system, see Marcus Clay and Dennis J. Blasko, "People Win Wars: The PLA Enlisted Force, and Other Related Matters," *War on the Rocks*, July 31, 2020; and Marcus Clay, Dennis J. Blasko, and Roderick Lee, "People Win Wars: A 2022 Reality Check on PLA Enlisted Force and Related Matters," *War on the Rocks*, August 12, 2022.

[49] Clay and Blasko, 2020; Clay, Blasko, and Lee, 2022.

[50] Congressional Budget Office, *The All-Volunteer Force: Issues and Performance*, 2007.

[51] Congressional Budget Office, 2007.

[52] James Burk, "Military Culture," in Lester R. Kurtz and Jennifer E. Turpin, eds., *Encyclopedia of Violence, Peace, and Conflict. Vol. 1,* Academic Press, 1999.

Applying the Logistics Culture Framework

Taken together, the factors outlined above provide a framework for understanding how organizational culture might positively or negatively affect a military's logistical capabilities. Various themes emerge from the framework. First, the effectiveness of a logistics organization is closely tied to the extent to which the broader military organization in which it is embedded prioritizes support functions relative to operational functions. Is sustainment built into the design of weapon systems, or is it something that is dealt with after the fact? Is adequate time allotted for routine maintenance?

Second, an organization's success is dependent on its ability to balance structure with autonomy. Standardized, ingrained routines and operating procedures enable organizations to effectively coordinate across multiple stakeholders in high-risk environments, such as those that are characteristic of military operations. At the same time, the complexity of logistics and maintenance functions indicates that every required action cannot be prescribed from the top; logisticians and maintainers must be empowered to use the resources at their disposal to identify the best solution for a given issue.

Third, an organization's primary function is to pursue its objectives within the context of scarce resources. There is no infinite supply of money, personnel, time, or equipment. At the same time, several of the logistics culture factors point to ways in which organizational culture might loosen the constraints imposed by scarcity. For example, does the organization invest in continual improvement through practices such as after-action reports? Which strategies are in place to leverage the latent capabilities of the civilian sector? How can workplace best practices, such as training, regular feedback, and nonmonetary incentives, contribute to a healthier, more productive workforce? In the Chapter 4, we apply these and other questions to the PLA.

Assessing the People's Liberation Army's Approach to Maintenance Management

In Chapter 2, we laid out the evolution of the PLA's approach to logistics and maintenance management, including how the PLA's organizational structure, activities, and policies have changed pre- and postreform. In this chapter, we seek to understand trends in how the PLA views maintenance as a factor of operational success and which efforts the organization prioritizes to mitigate weaknesses and strengthen capabilities. We identify the trends in approach that are shaped by the six maintenance success factors outlined in the Chapter 3 and characterize how those factors are incorporated into the PLA's maintenance organizational culture. This understanding provides insights into how capably the PLA might perform maintenance activities in practice.

In Chapter 2, we identified three motivating factors that underlie the PLA's efforts to reform: (1) lessons learned from logistics failures in previous conflicts, (2) an examination of Western militaries' approach to technological advancement, and (3) desires to realize greater efficiencies within the logistics system. This examination of logistics practices, particularly maintenance functions, suggests that the PLA appreciates the importance of maintenance as a factor of operational success.[1] There is an understanding that maintenance proficiency is both an indicator of technical skill within the force as well as a competency that might be lacking. Additionally, by

---

[1] This is substantiated by the way that the PLA views maintenance as critical for systems confrontation. See Engstrom, 2018.

studying other competitors' approaches, the PLA has identified areas to increase the efficiency of its maintenance practices. Viewed through the lens of the success factors from Chapter 3, these efforts indicate that the PLA has a culture of learning that balances studying other organizations with relying solely on its own experience.

The PLA has long demonstrated a culture of compensation when addressing the weaknesses that have been identified from previous conflicts. To address these weaknesses and bolster strengths, in recent years, the PLA has pursued a hybrid approach of compensation and improvement. Areas in which the PLA attempts to compensate for deficiencies include maintainer skill levels, lack of innovation, and poor knowledge-sharing. However, evolving technological advances in military equipment could outpace the PLA's maintenance capabilities, and a compensating approach might be tested under those conditions. There are also areas in which the PLA's approaches to maintenance improvements have led to gains in efficiency, such as process reforms and MCF. These improvements can lead to better operational outcomes, such as a reduced number of required maintainers, increased sortie generation, and higher levels of equipment availability for training and exercises.

A Note on Sources

In this chapter, we draw on a variety of Chinese sources, primarily Chinese-language articles published between 2018 and 2022 in the *PLA Daily* and republished on China's defense ministry website. To supplement these articles, we also drew upon professional military education texts, a wider array of PLA maintenance-focused articles published in the *PLA Daily* from 2008 to 2022, other Chinese state media and military sources such as the *People's Daily*, PLA service and TC news sites, provincial and regional news outlets, and select nonauthoritative Chinese media sources, such as Sina. Finally, we reviewed select English-language secondary sources that have assessed PLA logistics and maintenance practices, typically by Western scholars who research the PLA.

Many of the PLA and other Chinese-language media articles are geared toward motivating PLA audiences through narratives that identify challenges and depict maintainers and other PLA personnel as key problem-solvers. However, even as they paint the picture of "hero maintainers" who innovate to enhance the force's effectiveness, the articles identify several weaknesses, challenges, frustrations, and limitations faced by maintenance personnel. Like many public-facing sources on the PLA, unit details, equipment specifics, and geographic information are often omitted from these articles, which makes it difficult to tie specific practices to discrete unit types or make inferences about the maintenance of a given weapon system. Despite these limitations, the content provides useful insights into how maintainers operate by shedding light on developments within the organizational culture that shape their activities.

Lack of a Skilled, Professional Maintenance Force Within the People's Liberation Army

Maintenance proficiency is an indicator of the technical skill inherent in a military unit. Many factors impact maintenance proficiency, including individual knowledge and experience, knowledge-sharing and communication within teams, and the ability to innovate. PLA reforms recognize that maintenance skill levels are insufficient for the needs of the future battlefield because evolving technological advancements could outpace the PLA's maintenance capabilities. PLA sources indicate that there is a large divide between the skills of junior technicians and more-senior maintainers (who are predominantly senior NCOs), and limited programs have been developed to effectively train junior personnel to mitigate this weakness. Instead, the PLA has compensated for these weaknesses with technological solutions. This system leads to high reliance on skilled, senior maintainers who are often highly sought after by industry, but patchy skill levels at the novice and junior NCO level, where turnover is high. To address this weakness, PLA reforms include reducing divides between enlisted and officers and imbuing NCOs with more responsibility. However, these objectives have not yet been realized. Additionally, aspects of the PLA's maintenance organizational culture, such as hierarchical dynamics and risk aversion, are at odds with the PLA's intentions for reform.

The Officer-Enlisted Divide Affects Maintenance Units That Are Reliant on Conscript Forces

Social and cultural biases within the PLA create separation between the enlisted and officer ranks, making it difficult to fully realize the goal of professionalization of the enlisted force. Factors that affect this separation between the officer corps and NCOs include differences in compensation, trust issues, and quota systems that limit NCO participation on party committees; one PLA media source reflected these dynamics by characterizing NCOs as "afraid . . . [and] unwilling to take action."[2] Professionalization of the senior enlisted force is still a long way from being realized.

Reforms since the early 2000s have attempted to modernize the enlisted force in general by shortening the duration of conscription contracts, consolidating initial training at training centers, adjusting the conscription and demobilization schedule to occur more frequently, expanding the number of NCO ranks, attempting to imbue more responsibility into those ranks, and increasing enlisted pay and benefits.[3] These reforms have had some implications for maintenance units. Changes to where training resides for conscripts and to the demobilization schedules helped stabilize personnel in maintenance units, which heavily rely on conscripts. However, social biases that characterize enlisted as less than, along with shorter contracts for conscripts, make it difficult to retain quality personnel in maintenance teams. In Chapter 3, we identified that recruitment and retention are important aspects of a successful maintenance organizational culture. Lack of professionalization of the enlisted force coupled with a lack of professional maintenance culture might further compound issues related to recruitment and retention of a skilled technical labor force and might further affect unit readiness and operational capabilities.

---

[2] Clay and Blasko, 2020.

[3] Clay and Blasko, 2020. On the length of conscription, see Kenneth Allen, "The Evolution of the PLA's Enlisted Force: Conscription and Recruitment (Part One)," *China Brief*, Vol. 22, No. 1, January 14, 2022.

People's Liberation Army Reforms Have Given Noncommissioned Officers More Responsibility

The PLA recognizes that to meet the requirements of the future battlefield, it will need to increase the maintenance capability of *grassroots units* (those units at or below the battalion level), which hinges on improving the capability of novices and integrating them into maintenance activities.[4] It also recognizes the need for investment in the NCO ranks, giving NCOs more responsibility. For example, in the PLAAF, the separation of complex technical fixes from simple dispatch support helps gain efficiency while enabling NCOs to lead flight line maintenance, which allows officers to perform technical maintenance tasks elsewhere.[5] Similarly, the PLAAF employs rotating shifts of junior technicians before inspections by expert technicians.[6] This process effectively gives the novices some practical experience by enabling them to participate in maintenance tasks, but it continues to segregate junior maintainers from the more skilled senior technicians. It also potentially causes the junior maintainers to lose out on the ability to obtain greater skills through on-the-job training, which is an aspect that has been deemed important to ensure latitude of independent decisionmaking. Within the PLAA, there have been similar efforts to expand the overall maintenance capacity of the maintenance team to decrease dependence on elite maintainers. According to one article, previously,

> in order to ensure the integrity rate and dispatch rate of vehicle equipment, every time the equipment was overhauled and maintained, the grassroots battalion company would send the most capable technical backbone to battle, resulting in the repair experts being too busy.[7]

To address this practice, the brigade party committee mandated that novice maintainers attend skills training to limit idleness and increase proficiency.

---

[4] Zhang Z. and Huang K., 2022.

[5] Allen and Garafola, 2021, p. 270.

[6] Burke et al., 2016, p. 77.

[7] Che and Zhang, 2018.

Additionally, investment in professional maintenance education begins at the NCO rank. In the PLAAF, the primary institution in which this education occurs is the Air Force Engineering University's Aviation NCO School, which "over the past 70 years since its establishment . . . has successively trained and delivered more than 170,000 aviation maintenance personnel for the armed forces."[8] To ensure the currency of maintenance procedures taught, the NCO school has been sending instructors to grassroots units as part of an internship program to understand mishaps and maintenance issues that units face and provide technical support during exercises.[9] This practice has been widely used, with instructors traveling to at least ten different aviation brigades per year; each rotation lasts about two weeks and affords the instructors the opportunity to understand needed improvements to maintenance procedures based on interacting with the units while providing training to the soldiers.[10]

An Over-Reliance on "Hero Maintainers" Results in an Unbalanced Maintainer Force

Ministry of National Defense reporting highlights PLA dependence on elite maintainers at the expense of training for novice maintainers. Multiple human interest piece propaganda articles focus on Master Sergeant One and Two repair specialists (i.e., very senior enlisted members) and how their efforts ensure the operational readiness of the unit to which they are

---

[8] Xia Dong [夏冬] and Yang Fan [杨帆], "Air Force Engineering University Aviation Non-Commissioned Officer School Lifts the 'Wings' of the War Eagle to Take Off" [空军工程大学航空机务士官学校托举战鹰腾飞的"翅膀"], Xinhua, April 25, 2021.

[9] Zhang Yuqing [张玉清], Zhou Zhengxin [周正信], and Yang Fan [杨帆], "Air Force Engineering University Aviation Non-Commissioned Officer School Organizes Teachers to Go Fly with the Troops" [空军工程大学航空机务士官学校组织教员赴部队跟飞学习], Xinhua, August 15, 2019.

[10] Yang Fan [杨帆] and Li Jianwen [李建文], "Instructors from the Air Force Engineering University Aviation Non-Commissioned Officer School Arrive with the Troops to Follow the Flight Practice" [空军工程大学航空机务士官学校教员到部队跟飞实践], *PLA Daily*, August 30, 2020; Zhang, Zhou, and Yang, 2019.

assigned.[11] This reported level of knowledge contrasts with the low level of knowledge and skills to conduct basic maintenance on equipment among novices, making them reliant on technical specialists. As one example from the reporting illustrates, the most-capable repair experts could become extremely busy carrying out overhauls and repairs, while novice personnel were left "idle with nothing to do."[12]

Although these hero maintainers bring high levels of expertise to maintenance units, they are also frequently sought after by industry. The articles that highlight maintainer achievements commonly reference the number of times that they have been approached by industry but turned down the offers: "Over the past 24 years, he has devoted himself to the battle station, and three times has declined high-paying employment from local manufacturers."[13] Those senior enlisted personnel who do depart for industry positions are not described as "heroes" in PLA media, and the rate of departure is not reported.

Prior to beginning maintenance efforts in an organization, PLA recruits are trained in the basics and provided with technical data that cover the fundamentals of a weapon system. The more sophisticated maintenance knowledge, experience, and proficiency that are needed to excel are meant to be taught through on-the-job training by senior maintainers who pass on their knowledge, which differs from unit to unit if it occurs at all. The reporting does not indicate clear processes for this training, and it does not appear common that this knowledge is shared. For instance, one article references a senior maintainer who is about to time out of service. The maintainer compiles his knowledge into a book to share his experience with others, with the implication being that the knowledge had not been shared previously:

---

[11] The "elite" maintainers are listed as Master Sergeant Class One [一级军士长] or Master Sergeant Class Two [二级军士长] (Tang Zhiyong [唐志勇] and Chen Wanjin [陈万金], "Retired Veteran Xu Daming: Repairing Boats for 30 Years; Ensured Nearly 1,000 Sailings with Zero Mistakes" [退役老兵徐达明:维修船艇30年，近千次保障出航零失误], *PLA Daily*, December 8, 2020). Note that Xu served in the People's Armed Police but held a similar rank level of Chief Sergeant Class One [一级警士长] (Chen and Huang, 2019; Tang and Wu, 2019). On PLA NCO ranks, see Clay, Blasko, and Lee, 2022.

[12] Che and Zhang, 2018.

[13] Chen and Huang, 2019.

> Zhan Wu has another job, which is to organize more than a dozen notebooks into one volume. This is because in a few years he will reach the maximum number of years of service as a soldier. Zhan Wu said: "I didn't do anything earth-shattering, I just wanted to pass on the experience and technology I accumulated over 20-plus years to contribute to the reform and strengthening of the military."[14]

Multiple articles focus on individual pursuits rather than the success of a unit. This focus on the individual gives the impression that, at the lower tactical levels, more repairs are being done locally than they be would in typical U.S. maintenance units.[15] The focus on the individual hero maintainer and references to limited knowledge-sharing create a maintenance culture in which information resides with individuals and is siloed within units rather than being shared more broadly across the organization. In Chapter 3, we identified that a critical question to determine the maturity of an organization's culture of learning is to assess whether continuous learning and improvement processes related to maintenance have been institutionalized. These examples point to less institutionalization and more ad hoc, decentralized processes.

Lack of Pride in the Maintenance Profession Affects Attention to Detail and Independent Judgment

Across the PLAA and the PLAAF, there appears to be a lack of professional pride and attention to detail in supporting specialties.[16] For example, in an article about a soldier who is responsible for implementing quality assurance processes, the soldier describes routinely observing "mistakes and omissions" during his inspection of maintenance spaces: "[s]mall things

---

[14] Jia Baohua [贾保华], Yang Lei [杨磊], and Xiang Shuangxi [相双喜], "Compiling an 'Encyclopedia' for Equipment Maintenance" [为装备维修编制"百科全书"], *PLA Daily*, April 5, 2018.

[15] For an example within the PLAAF, see Zhang, Zhang, and Feng, 2019.

[16] For examples of these dynamics, see Burke et al., 2016.

such as rags strewn about, cloth fabric not stacked neatly, and extra items in the toolbox."[17]

Soldiers and officers take an outdated view of support services, believing that support services are in some respects lesser than those specialties that conduct more-traditional warfighting responsibilities, such as pilots. This perspective results in a lack of a formal professional culture in maintenance units, which manifests as maintainers adopting a formal adherence to procedures rather than implementing procedures voluntarily. In what appears to be an attempt to counter this bias, propaganda pieces emphasize narratives that highlight pride in the maintenance profession. For example, one piece describes an air force mechanic having personal responsibility for the aircraft he maintains: "This is the pilot's fighter jet, and it is also his [the maintainer's] fighter jet."[18]

A lack of pride in the profession also leads to a tendency for maintainers to lack attention to detail. Articles that detail quality assurance processes include examples of procedure-cutting practices that affect the quality of maintenance and jeopardize safety. Quality inspectors within the PLA blame safety issues on carelessness and inability to attend to small but necessary maintenance tasks, such as tool accountability.[19] Compounding this issue is the high priority that the organization places on error-free maintenance work. Many of the PLA articles contain similar headlines that highlight a specific maintainer and how many hours of work they have performed or how long equipment has run with no errors.[20] Maintenance organizations that emphasize a zero-defect mentality are at risk of corruption and cover-ups to maintain the appearance of a perfect record. Despite reforms to address these problems, however, PLA maintainers do not appear to be wholly accepting when quality assurance procedures are addressed,

---

[17] "As the 'Gatekeeper' of Aircraft Safety, He Is Willing to Be the 'Bad Guy'" [作为机务安全"把关人", 他甘当"坏人"], *China Military Online* [中国军网], March 30, 2018.

[18] Li et al., 2021.

[19] "As the 'Gatekeeper' of Aircraft Safety, He Is Willing to Be the 'Bad Guy,'" 2018.

[20] Examples include Tang and Chen, 2020; and Tang and Wu, 2019.

complaining that the processes are "annoying" and that quality inspectors are seen as the "bad guy."[21]

Within the PLA, the latitude for independent decisionmaking appears low because of risk aversion, lack of professional pride within maintenance units, and lack of appropriate skill levels at lower levels of the force. Independent decisionmaking is further impeded because maintenance management might support the objectives of the controlling mechanism rather than the needs of the maintenance-performing organization. At the highest levels, General Secretary and CMC Chairman Xi Jinping has prioritized innovation across the defense enterprise to move the PLA from being "an innovation follower to becoming an original innovation leader."[22] Reporting indicates that this top-down prioritization of innovation is being pushed by military leadership as well, with an emphasis on training innovation, but risk aversion among soldiers and officers impedes progress. For example:

> some officers and soldiers are worried about the risks of innovation, and think that more things are worse than less things; some officers and soldiers are worried that innovation-related failures will lead to criticism from the people around them, and as a result they hesitate . . . . This not daring to make mistakes and can't afford to be wrong type of mentality has shackled many people's hands and feet, bringing their footsteps to a halt when facing innovation.[23]

Independent thinking is also hindered by the lack of knowledge, skills, and abilities of the average maintainer in the PLAA and PLAAF. Maintainers have the ability to understand that a maintenance problem exists but lack the skill level to solve that problem. One article noted that "facing the

---

[21] "As the 'Gatekeeper' of Aircraft Safety, He Is Willing to Be the 'Bad Guy,'" 2018.

[22] Tai Ming Cheung, "Keeping Up with the *Jundui*: Reforming the Chinese Defense Acquisition, Technology, and Industrial System," in Philip C. Saunders, Arthur S. Ding, Andrew Scobell, Andrew N. D. Yang, and Joel Wuthnow, eds., *Chairman Xi Remakes the PLA: Assessing Chinese Military Reforms*, National Defense University Press, 2019, p. 591.

[23] Xu, Sun, and Xing, 2021.

dilemma of new equipment breakdowns, many people have insufficient basic knowledge, and it is difficult to flexibly use and integrate them."[24]

There are some indications of innovation at the grassroots level. Initiatives previously mentioned, such as expert studios to exchange best practices and technology solutions (e.g., equipment maintenance records), demonstrate that individual maintainers are coming up with local solutions to fix maintenance challenges. However, these are highlighted in the reporting as singular events that depend on the hero maintainer and do not seem to result from organizational culture. Therefore, to answer a question raised in Chapter 3, there appear to be few incentives for maintenance technicians to exercise independent thinking and judgment.

To Compensate for Lack of Skill and Improve Maintenance Efficiency, the People's Liberation Army Prioritizes Process Reforms and Technology Investment

PLA maintenance reforms have heavily prioritized process improvement and technology solutions. The routinization of maintenance functions is important within an organization because routinization ensures a shared understanding of the division of labor and hierarchies while enabling individuals to implement a standard approach to maintenance functions. A review of the relevant literature indicates that the PLA prioritizes the routinization of maintenance within its organizations to create established and well-understood activity hierarchies. The PLA has worked to better prioritize routinization of maintenance within its organizations through such improvements as changes to the frequency of maintenance tasks and the delineation of labor within maintenance sections. These efforts have led to improvements in efficiency and operational gains. The PLA has also instituted changes to its quality control procedures, which center on clear hierarchies and division of labor, to overcome issues with cover-ups and corruption. Similarly, the PLA has invested in technology solutions to mitigate low

---

[24] Li, Xing, and Lei, 2019.

skill levels of novice and junior maintainers, although these solutions do not appear to be broadly institutionalized. Technology solutions alone, however, are not able to overcome the organizational cultural biases outlined in the previous section to institute meaningful change.

Improvements in Routinization of Maintenance Have Led to Greater Efficiencies and Operational Outcomes

Maintenance tasks within the PLAA broadly fall along a two-tier hierarchy, which is similar to U.S. military maintenance structures. This hierarchy of activities was presented in Table 2.2. PLAA repair companies in combined arms battalions or armor brigades routinely conduct troubleshooting, equipment inventory, and some emergency maintenance. These light-level repairs correspond to field-level maintenance in the U.S. context, according to U.S. military maintenance publications.[25] Sustainment-level maintenance tasks are relocated from the operational unit and regularly performed at a qualified repair unit or by the manufacturer. For the PLA, these tasks include routine diagnostic checks, equipment modification, equipment acceptance checks, and postmission service determined by system hours of operation or mileage.

Within the PLAAF, maintenance support teams at unit repair shops are organized to include a flight line and a workshop. The PLAAF refers to these as level 1 and 2 repairs, which take place at unit repair shops; level 3 repairs are handled at aviation repair facilities.[26] The flight line is responsible for clearing and testing aircraft; the workshop contains the most-skilled technicians who conduct technically challenging maintenance tasks. This division of labor reform reduces the number of required maintenance personnel; one person is needed to do a job that previously required five people.[27] Similar reforms were enacted in 2017 in an Eastern TC Air Force (TCAF) air brigade to address the same issues. The delineation of tasks between the flight line and workshop separates simple dispatch support from complex techni-

---

[25] Army Technical Publication, *ATP 4-33: Maintenance Operations*, Department of the Army, July 2019.

[26] Allen and Garafola, 2021, p. 264.

[27] Burke et al., 2016, p. 78.

cal fixes, which the PLAAF believes will save time and money through efficiency gains. For instance, the Eastern TCAF air brigade reported reducing engine inspection and maintenance times by 130 minutes per engine.[28]

For the timing of maintenance activities, the PLA has instituted reforms that move away from more rigid, calendar-based scheduling to a more flexible, flight hours–based approach. Some of these changes were implemented more than a decade ago. In 2011, a Nanjing Military Region Air Force unit instituted a maintenance schedule determined by flight hours; the previous schedule required the unit to conduct down days twice a year. This grouping of maintenance schedule by aircraft flight hours prevents the grounding of the entire fleet at the same time.[29] Other units have also experimented with precision and lifetime maintenance models that prioritize efficiency and inspections across the lifetime of the aircraft over replacing parts according to prescribed time intervals in preventive maintenance policies.[30] This routinization of maintenance demonstrates a commitment to process improvement rather than compensation to address deficiencies.

PLA ground force reforms in 2015–2016 attempted to further address maintenance efficiency issues by transitioning from organizing maintenance repair teams by weapon systems (e.g., armor, ordnance, vehicles) to organizing repair teams by objects (e.g., chassis, bodywork, special equipment).[31] The intention was to improve the comprehensiveness of maintenance support and the standardization of maintenance procedures and to increase efficiency. A move from consolidation to specialization is interesting because of the already identified deficiencies in skill levels of the maintainer workforce. Studies of the U.S. Air Force show that, in the

---

[28] Allen and Garafola, 2021, p. 270.

[29] Burke et al., 2016, p. 77.

[30] Allen and Garafola, 2021, p. 270.

[31] Liu Tielin [刘铁林], Wu Yongle [武永乐], Zhang Haifeng [张海峰], and Huang Xinxin [黄欣鑫], *Army Unit-Level Equipment Maintenance Support Operation Mode* [陆军部队级装备维修保障作业模式], National Defense Industry Press [国防工业出版社], 2022, pp. 140–142; Wang and Cheng, 2022.

long run, consolidation can lead to more efficiency in use of manpower and reduction of costs.[32]

Quality Assurance Processes Ensure Maintenance and Inspection Standards

The PLA has recognized deficiencies in its quality assurance practices that require attention. The PLAAF's early standard of practice was that individuals who conducted maintenance also performed quality control inspections on the equipment. This practice was acknowledged by the PLAAF to result in cover-ups and lead to poor quality control.[33] Reforms from 2010 to 2011 created a separation between maintenance and inspection teams. Inspection teams created rules, regulations, and procedures for quality assurance processes that were independent of maintenance practices. The reforms resulted in "maintenance management offices" and "safety supervision offices" in the maintenance shops that have been replicated across the PLAAF.[34]

Quality inspection reforms through workforce design appear to be continuing. In the Northern TCAF Support Department's field support office, a talent team was constructed and an aviation maintenance technology research room was created to closely observe combat training support, including maintenance functions.[35] The intent of these changes is to improve quality and efficiency of services. The talent team visits the maintenance team quarterly to inspect, teach, and conduct research on maintenance issues.[36] Regulations require a "card-holding" technique to ensure quality control; the technique consists of "one person reading the card,

---

[32] Thomas Light, Daniel M. Romano, Michael Kennedy, Caolionn O'Connell, and Sean Bednarz, *Consolidating Air Force Maintenance Occupational Specialties,* RAND Corporation, RR-A1307-AF, 2016.

[33] Burke et al., 2016, p. 78.

[34] Burke et al., 2016, p. 79.

[35] "Keep an Eye on Combat Training Support and Cultivating Scientific and Technological Talents" [紧盯战训保障 培育科技人才], *PLA Daily* [解放军报], June 7, 2022.

[36] "Keep an Eye on Combat Training Support and Cultivating Scientific and Technological Talents," 2022.

one person operating, and one person inspecting."[37] Reporting also references safety issue bulletins and lists maintained in units, and, when safety issues appear, there are mechanisms to determine whether the problem is persistent across the aircraft regiment.[38] While these changes are not fully assimilated into the workplace culture, they do indicate a routinization of maintenance functions and that quality control procedures and routine maintenance inspections are required, implemented, and documented on a regular basis.

Investment in New Technology and Processes Compensates for Maintenance Deficiencies Rather Than Improving Them

PLA texts place heavy emphasis on informatization and investment in technology to advance maintenance approaches. This emphasis includes encouraging the integration of technology systems into maintenance support processes, applying new support technologies to facilitate maintenance practices, and the use of virtual platforms for maintenance training.[39] The technology solutions compensate for poor skill levels, but technology solutions alone might not lead to wholesale improvements across the force.

In the PLAA, a primary innovation has been a move from paper-based recordkeeping to data-driven solutions. Several sources allude to the introduction of an equipment "medical record" (病史) or a "digital maintenance file" across the PLAA that would greatly improve automated record keeping.[40] However, some sources still highlight the maintainers' use of personal logbooks as sources of documentation for previous repairs, which indicates uneven implementation across units. Another source highlights the use of a "repair station machine" to improve maintenance practices, including uploading pictures, videos, and specifics about the repair process into a

---

[37] "As the 'Gatekeeper' of Aircraft Safety, He Is Willing to Be the 'Bad Guy,'" 2018.

[38] "As the 'Gatekeeper' of Aircraft Safety, He Is Willing to Be the 'Bad Guy,'" 2018.

[39] Shu Zhengping [舒正平], ed., *Science of Military Equipment Maintenance Support* [军事装备维修保障学], National Defense Industry Press [国防工业出版社], 2013, pp. 214, 247. On integrating technology, see Luce and Richter, 2019.

[40] Chen, Feng, and Zhou, 2022; Guo and Hai, 2021.

system when a repair task is completed. In one unit this has been performed for more than 100 different types of repairs.[41]

Similar investments have been made in information technology for the PLAAF. Beginning in 2011, PLAAF units have been documented as moving from paper-based records to network solutions.[42] Maintenance units have instituted network-driven solutions to display real-time data on aircraft conditions (airframes, engines, and others). Flight lines also use handheld devices to relay aircraft conditions to the diagnostic center to perform systems analyses. The devices also contain troubleshooting information.[43] Although these maintenance networks have been created, the overall integration of information technology is unclear because there is limited discussion of these changes across dissimilar units. Either the changes are deemed too sensitive to discuss or these efforts served as boutique solutions that were not broadly integrated internally or externally. According to the logistics success factors, technology can enable the improvement of maintenance processes from a culture-of-learning angle. However, the technologies pursued by the PLA appear to be insufficient to fully overcome other organizational impediments, such as the lack of institutionalized processes.

Reforms Aimed at Maintenance Standardization Are Not Institutionalized

Within the PLA there is a proclivity to strictly adhere to hierarchies within the rank structure, party loyalty, or even military specialties (such as combat arms versus service support). Despite this adherence to strict hierarchies, the PLA's propensity for not sharing information in and across maintenance units tends to dominate and lead to ad hoc maintenance practices rather than institutionalized processes. Additionally, top-down dictation of maintenance procedures creates inefficiencies in the system and increases maintenance times.

---

[41] Guo and Hai, 2021.

[42] Allen and Garafola, 2021, p. 268.

[43] Burke et al., 2016, p. 76.

People's Liberation Army Directives Emphasize a Need for Standardized Maintenance Procedures

PLA texts emphasize strengthening the standardization of maintenance procedures so that all repair units can perform to the same technical standards, and those technical standards will drive improvements to product quality.[44] These standards are nested hierarchically, with the highest level being the National Military Standards (国家军用标准 or *guojia junyong biaozhun* [GJB]). GJB maintenance manuals are the explicit forms of national maintenance programs for military-grade equipment. GJB manuals might not be used with high regularity, but such documents as the *Requirements and Methods for Formulating Preventive Maintenance Program of Equipment* [装备预防性维修大纲的制定要求与方法] were in circulation through 2012 and were meant to routinize maintenance best practice across the PLA.[45] The services maintain their own service-specific plans, in addition to the GJB, as outlined in Table 4.1.

The origins of maintenance procedures do seem to have changed. Previously, the procedures were compiled and issued by the frontline units, but, as the systems have evolved, they are written by the factory where the systems are produced. This shift has led to some reported issues of mainte-

TABLE 4.1
Service-Specific Military Standards or Maintenance References

PLA Service	Standard
PLAA	Equipment Maintenance Manuals (装备维修手册), as noted in Ministry of Defense and PLA press reporting[a]
PLAN	Naval Military Use Standards (海军军用标准)[b]
PLAAF	Air Force Military Standards (空军军用标准)[c]

NOTE: Pre-2015 sources discuss the ground forces, but the PLAA and PLARF officially became independent services in 2015. We did not find a distinct term used for the PLARF.

[a] Zhang Taiyao [张泰耀] and Huang Yongjian [黄永健], "Troubleshooting Introduces 'Cloud Maintenance'" [故障处置引入"云端维修"], *PLA Daily*, October 30, 2022.

[b] Shu, 2013, p. 136.

[c] Zheng Dongliang [郑东良], ed., *Aviation Maintenance Theory* [航空维修理论], National Defense Industry Press [国防工业出版社], 2012.

---

[44] Shu, 2013, p. 98.

[45] For more on the GJB systems, see Shu, 2013, p. 136.

nance standards that do not match the ground truth at the maintenance unit, which leads to time-intensive coordination meetings with division and regimental leaders, maintenance personnel, and factory representatives.[46]

In Reality, Ad Hoc Maintenance Practices Dominate

Notably, many propaganda pieces featured senior maintenance NCOs identifying ad hoc maintenance solutions, which perhaps suggests a lack of standardization across service branches and TCs. Although PLA reporting advances the importance of self-study and deference to general maintenance PLA manuals, such as *Equipment Common Faults and Troubleshooting Methods* (装备常见故障及排除方法), and more platform-centric maintenance manuals, such as *New Armored Equipment Command and Communications Repair Guide* (新型装甲装备指控通信修理指南), these reports also make explicit reference to NCOs finding clever workarounds to problems or equipment faults that are not mentioned in existing PLA maintenance publications. While these workarounds are evidence of innovative solutions at the tactical level, they are not broadly institutionalized. For example, PLA maintenance cadres are deputized to publish maintenance circulars (维修小报) or corrections to existing maintenance SOPs. These circulars were not reported as being disseminated outside individual TCs, which provides additional evidence that PLA maintenance standardization is limited by theater geographies.

Specific to the ground forces, media reporting indicates that the numbered Group Armies, which are corps-level organizations, separately contend with their maintenance strategies by undertaking outreach and coordination unique to the Group Army or its subordinate units. For example, one article describes a brigade in the 72nd Group Army as using "self-innovated" maintenance processes to improve battlefield support.[47] Local units invest time working with universities and academic institutions to

---

[46] "As the 'Gatekeeper' of Aircraft Safety, He Is Willing to Be the 'Bad Guy,'" 2018.

[47] Gong Peiwen [巩沛文] and Tong Zujing [童祖静], "A Brigade of the 72nd Group Army Uses Self-Renovated Maintenance Equipment to Improve the Level of Battlefield Support" [第72集团军某旅运用自主革新的维修装备，提升战场保障水平], *PLA Daily* [解放军报], September 13, 2021.

find solutions to equipment maintenance issues, such as issues with flotation assistance equipment, but it is not clear that the innovations are shared with other similar brigades or that the analysis is systematically applied.[48] Similarly, in a brigade of the 77th Group Army, a team created an innovative device to fix issues with towing and rescue and implemented "expert workshops" to share information within the brigade because there was no existing mechanism for mutual learning and collaboration.[49] In the article, there is no discussion of broader communication of these practices. Another article highlights that technicians understand the importance of learning from past mistakes but reference reading through their previous handwritten logs rather than referring to the service manuals that are distributed to units that field the same equipment.[50]

In the PLAAF, knowledge-sharing of maintenance procedures might be more institutionalized. Instances of maintainers writing "repair process regulations" when determining a fix for a new technology have been documented.[51] For one "new bomber" type, a maintenance schoolhouse instructor found disconnects between teaching materials and the technical specifications required for the aircraft, so he compiled new materials to create a new "aircraft construction and maintenance" textbook.[52] Although this demonstrates some degree of documentation, it is not mentioned how broadly these regulations are promulgated or read.

Additionally, there are hints that ad hoc maintenance needs are a product of the PLA's budgetary abundance in recent years. As military modernization occurred, air and ground platforms were sometimes mentioned as being too numerous for any single cadre to manage. Sufficiency of time to adequately maintain equipment might be compromised in units that are

---

[48] Gong and Tong, 2021.

[49] Xu, Sun, and Xing, 2021.

[50] Jia, Yang, and Xiang, 2018.

[51] Shi Feng [石峰] and Guo Rong [郭蓉], "Repair Is Like Playing Hide and Seek with Faults, but He Has 'Piercing Eyes'" [维修就像是与故障玩捉迷藏，可他有"火眼金睛"], *PLA Daily*, April 17, 2020.

[52] Xia and Yang, 2021.

required to maintain a wide array of weapon systems. For example, one unit's technical repair team was assigned to manage and maintain

> over 40 types of tanks, armored vehicles, artillery, and other equipment, totaling more than 300 sets . . . . [These included a mix of] scrapped, decommissioned, in-service, and not finalized [equipment], spanning more than 60 years [of service].[53]

Maintaining multiple types of equipment at various levels of service is time intensive and limits maintenance shops from achieving economies of scale in their maintenance processes. Sources also note insufficient time to master new technology. When new systems are fielded to the units, maintainers lack sufficient time to learn maintenance of systems before the equipment is exercised.[54]

The cumulative reliance on ad hoc maintenance solutions, PLA maintenance circulars (which are apparently self-published by grassroots maintenance units), and other stories of maintenance improvisation at different echelons are as much evidence of a lack of maintenance standardization as they are proof of operational flexibility. The variety of PLA maintenance activities also hints at the variety of roles that maintenance cadres perform across echelons and throughout the yearly calendar. In-garrison, there is training and some degree of process improvement emphasized while out-of-garrison there are service and theaterwide exercises. There is year-round operational support to repair and maintain equipment as it ages and degrades from use.

Top-Down Maintenance Planning Dominates, Which Can Create Bureaucratic Processes That Prohibit Speed of Decisionmaking

In organizations in which the issuing of technical directives and routine maintenance planning resides with higher authorities, there might be a lack

---

[53] Liu, 2022.

[54] Lei, Wen, and Li, 2021.

of autonomy. This dynamic can lead to bureaucracy that slows down processes and limits decisionmaking authority at lower levels.

There is some evidence that this dynamic exists in the PLA. Multiple articles reference that the brigade party committee has the authority to institute changes to the maintenance procedures, such as dictating implementation of new initiatives at grassroots levels, assigning personnel, and approving certain technical fixes, at the battalion and company levels.[55] For example, as new service support battalions stand up in the PLAA, the brigade party committee directly controls the assignment of individuals and has a direct say in training:

> The party committee of the brigade ensures to the greatest extent possible that the cadres of the service support battalion are assigned to professional counterparts, and regularly organizes the cadres of the service support battalion to carry out the activities of "small training within the battalion and large training with the [broader] organization" to provide them with a good environment and sufficient "nutrients" for their growth.[56]

Approval for certain maintenance fixes is retained further up the chain of command, adding time to the process and centralizing decisionmaking further up in the organization. Superiors are responsible for allocating materials for maintenance and are often criticized for taking too long if the materials are even procured.[57] For repairs that require evacuation to the factory or manufacturer, coordination between the unit and the factory takes time to achieve, and the back and forth is assessed to "inevitably affect the training of the troops."[58] Some maintainers, not trusting the allocation system, purchase materials required for maintenance tasks them-

---

[55] Che and Zhang, 2018; "Eight 'Innovation Studios' Named After Officers and Soldiers" [8个"创新工作室"以官兵名字命名], *PLA Daily* [解放军报], March 2, 2020; Xu, Sun, and Xing, 2021.

[56] "The 'Three Questions' of Professional Training in the Service Support Battalion" [勤务保障营的专业训练"三问"], *PLA Daily* [解放军报], April 28, 2020.

[57] Tang and Wu, 2019.

[58] Lei, Wen, and Li, 2021.

selves.[59] In the previous chapter, we identified that a critical component related to the routinization of maintenance is whether maintenance groups are provided with sufficient time and resources needed to effectively carry out maintenance work. These examples indicate that certain processes might affect the sufficiency of time, but assessing the prevalence of these issues and their overall impact is difficult within the scope of our methodological approach.

Military-Civil Fusion Has Enhanced Maintenance Interoperability; Lack of Knowledge-Sharing Impedes Effectiveness

Similar to standardization of maintenance procedures, tensions between top-down directives and lack of knowledge-sharing affect MCF efforts.[60] The PLA recognizes that, as weapon systems become increasingly complex, its ability to rely solely on military maintainers becomes more strained. To overcome this challenge, the PLA highlights the need to strengthen MCF (earlier conceptualized by the PLA as military-civilian integration) to enhance maintenance support. This need is viewed as a strategic issue. In 2007, the CMC proposed to "further study and deepen the maintenance and support of military-civilian integration equipment" during the Eleventh Five-Year Plan.[61] In 2010, the CMC emphasized the reform as one of the PLA's "main points."[62] In 2011, the "All-Military Equipment Work Instructions" proposed the implementation of maintenance pilot projects.[63] These projects have been deemed by the PLA to have laid a foundation for further reform.

---

[59] Li et al., 2021.

[60] On the development and implications of MCF writ large, see Elsa B. Kania and Lorand Laskai, "Myths and Realities of China's Military-Civil Fusion Strategy," Center for a New American Security, 2021.

[61] Shu, 2013, p. 248.

[62] Shu, 2013, p. 248.

[63] Shu, 2013, pp. 248–249.

An example of military-civilian innovation in practice occurred during Mission Action 2010, when the 127th Light Mechanized Infantry Division of the 54th Group Army in the then–Jinan military region demonstrated the ability to move over a large distance (1,000 km along the Qingdao-Yinchuan Expressway) by leveraging military-civilian logistics support:

> They did not repair vehicles nor cook meals by themselves, and completely relied on the support forces of friendly military units and local civilian institutions along the route of movement. As a result, the support time was shortened by half . . . . The division sorted out six categories of 38 mobile support nodes, arranged collaboration with the friendly units and the local civilian support resources, and established a network of precise and efficient support along the route of movement. They also arranged reserved services with main repair and maintenance factories along the route of movement, shaping an emergency repair support system for more than 30 types of vehicles in the whole division.[64]

Top-Down Efforts to Reform Military-Civil Fusion Have Expanded to Include Maintenance Services

Past reviews of the PRC's national strategy of MCF, led by the Central Military-Civilian Fusion Development Committee (中央军民融合发展委员会), have emphasized the ways in which military equipment could be provisioned to the PLA under the principle of "one input has two outputs" (一份投入、两份产出).[65] However, MCF extends to both equipment and services that the civilian or commercial sector could provide. PRC guidance on MCF does not typically highlight maintenance topics—Chinese sources also seem to emphasize equipment over services—but PLA sources on maintenance best practices do draw a direct connection between MCF efforts the and

---

[64] "Jinan MR Infantry Division Relies on Local Support in Long-Range Maneuver," *PLA Daily Online*, October 24, 2010.

[65] Han Qiulu [韩秋露], Kong Zhaojun [孔昭君], Deng Xiaotong [邓晓童], "Analysis of National Defense Mobilization Construction Under the Background of Military-Civilian Fusion" [军民融合背景下的国防动员建设分析], *China Economic and Trade Herald* [中国经贸导刊], No. 3, March 2019.

the possibility of leveraging civilian talent to address maintenance needs. As previously mentioned, a 2013 source states that in 2011 military equipment guidance given to the PLA, there were formal steps taken to support or authorize MCF pilot projects or tests:

> In the "All-Military Equipment Work Instructions" [全军装备工作指示], it was proposed that "in accordance with the reform plan of the military-civilian integrated equipment maintenance support system, [the PLA] launch pilot projects for general equipment maintenance, specialized equipment overhaul and mutual support, and military-selected civilian equipment socialized maintenance support, in order to deeply explore the development path of military-civilian integration."[66]

From 2011 to 2018, PLA maintenance sources suggest that these MCF pilots have morphed into some level of maintenance best practice; MCF contracting options are now a standard consideration for military equipment managers. A 2018 guide to on-site managers of military equipment describes their role in MCF as

> utiliz[ing] the enterprise support platform and network to jointly launch a forward- and backward-extending MCF integrated support system [leveraging] resource sharing and complementary advantages with equipment manufacturers and scientific research institutes, to form a comprehensive support capability system for new and large-scale specialty equipment.[67]

By 2021, it seemed that MCF maintenance was one of several "modes" (模式) a military equipment manager is supposed to be aware of. Under a section titled "Selecting Maintenance Modes," one source states that

> [t]he maintenance and support modes of equipment mainly include the military's independent support, the full-life cycle support of the

---

[66] Shu, 2013, pp. 248–249.

[67] China Association for Quality [中国质量协会], ed., *On-Site Management Guide for Military Equipment Maintenance and Support Enterprises* [军用装备维修保障企业现场管理指南], China Standards Press [中国标准出版社], 2018, p. 7.

research and development units, and the military-civilian fusion support, etc. When formulating the equipment maintenance support plan, the characteristics of the support object, the military and civilian support capabilities, the maintenance support requirements and the funding needs should be comprehensively considered.[68]

MCF entails equipment and services, but a heavier emphasis had been placed on equipment than on services. More recently, it appears that MCF is considered a legitimate option for maintenance services and not simply an experiment. MCF does provide a unique pathway for the PLA to draw expertise, manpower, and other resources from the civilian logistics sector, which demonstrates a certain level of agility in leveraging the civilian sector. However, it is difficult to assess how widely used MCF is for maintenance services. This is an area for further study.

Reforms to Life-Cycle Management Appear Promising to Improve System Maintenance and Reliability

The PLA has studied the United States' and other Western countries' approach to LCM. Areas of focus include the importance of engineering for reliability and maintainability of systems during the development, construction, and continued support of equipment.[69] The PLA maintains a systematic defense LCM approach that contains specific phases, consisting of proposal, design, engineering and development, finalization, and batch production.[70] Although the overall approach is complex and involves various

---

[68] Jiang Tiejun [蒋铁军] and Zhang Huaiqiang [张怀强], *Ship Equipment Support Engineering Series: Equipment Maintenance Support Planning Management* [舰船装备保障工程丛书–装备维修保障计划管理], China Science Publishing [科学出版社], 2021, pp. 124, 270.

[69] Shu, 2013, p. 244.

[70] Mark Ashby, Caolionn O'Connell, Edward Geist, Jair Aguirre, Christian Curriden, and Jonathan Fujiwara, *Defense Acquisition in Russia and China*, RAND Corporation, RR-A113-1, 2021, pp. 17–18.

actors and processes, we identified some strengths and weaknesses of the approach from a maintenance perspective.

One weakness in the approach is appropriate oversight of the process. Prior to the 2015–2016 reforms, the GAD was the critical military player providing oversight of the process, but that oversight primarily applied to the ground forces, with limited oversight of the other services. This discrepancy resulted in the nonground services overseeing their own LCM, which increased variability and decreased standardization in the process.[71] Variation and nonstandard oversight can affect system quality and production times and lead to outdated technology or early obsolescence of weapon systems.[72] Recent reforms have reorganized the GAD into the Equipment Development Department (EDD) under the CMC. LCM oversight is under the purview of the EDD, but the reforms stripped the department of its oversight of other key actors in the system, particularly the GAD Science and Technology Committee (now the CMC Science and Technology Committee). The evaluation of these reforms to the LCM process warrants further analysis, but some researchers assess that, in the long term, these reforms could reduce inefficiencies and corruption in the system.[73] The changes could lead to greater oversight across the force and increase reliability and maintainability. Additionally, the EDD's centralization of oversight across the force could lead to greater joint approaches to LCM and eliminate earlier biases under the GAD toward the ground force.[74]

Another weakness is the PLA research, design, and acquisition process, which is highly bureaucratic and lengthy. Significant amounts of time and financial resources are invested in engineering phases that can quickly reach low-rate initial production but require a significant amount of time to reach full-rate production because of quality control issues.[75] Once fielded

---

[71] Puska et al., 2014.

[72] Puska et al., 2014.

[73] Joel Wuthnow and Phillip C. Saunders, *Chinese Military Reforms in the Age of Xi Jinping: Drivers, Challenges, and Implications*, Institute for National Strategic Studies, National Defense University, March 2017, p. 36.

[74] Cheung, 2019, p. 592.

[75] Puska et al., 2014.

to a unit, there can be other time delays before systems reach initial operational capability because of the unit starting from scratch to train, learn maintenance activities, and write new regulations for operations and maintenance.[76] One source notes that these efforts can even take longer if the system comes from another country, such as Russia, because of translation requirements and the fact that the PLA does not always receive maintenance and operating manuals from the exporting country.[77]

Although the PLA's research, design, and acquisition process takes a considerable amount of time, there are some strengths in the system from a maintenance perspective. At least for certain services, military officers provide operational input to the design and engineering phases of the system development, which includes perspectives on maintenance procedures. This has been evidenced in the PLARF, where officers from seed units provide input to the factories and research institutions that are responsible for the development of the equipment. The seed units would then be upgraded to fully operational units once the weapon system was fielded.[78] Similar procedures are referenced in PLAAF units, which are generally centered on one specific weapon system type.

Military representatives at equipment factories are tasked with the oversight of system requirements during the design and engineering phases. In theory, this provides military input to ensure standardization, reliability, and interoperability. In practice, there are weaknesses in this relationship because the military representatives are paid by the firms, not the military, which affects their impartiality, and they might lack the technical expertise required to provide appropriate insight.[79] To determine whether these weaknesses in the PLA's LCM system might result in maintainability and reliability issues, we conducted a case study of the PLA's export of Cai Hong-4 (CH-4) systems.

---

[76] Puska et al., 2014.

[77] Allen and Garafola, 2021, p. 348.

[78] Ashby et al., 2021, p. 18.

[79] Puska et al., 2014.

Proxy Analysis Excursion: International Customers' Experience Maintaining China's CH-4 Unmanned Aerial Vehicles

Over the course of our research, we investigated equipment that is exported to foreign militaries by China and fielded by the PLA to determine whether we could gain insights from maintenance issues that affect those pieces of equipment and might further illuminate PLA maintenance issues. To identify potential candidates, we reviewed the literature to determine which types of equipment that China exports were also most often referenced in PLA maintenance articles. This equipment included artillery, armor, unmanned aerial vehicles (UAVs), and fighter jets. We were able to further reduce the number of platforms by determining whether China exported the same variant of the equipment that the PLA itself employs. This process produced a few possible candidates. The CH-4 Rainbow (彩虹) medium-altitude, long-endurance drone was the most promising because Chinese exports it to Indonesia, Iraq, Jordan, Laos, Nigeria, and Saudi Arabia.[80] Transfers to multiple foreign customers provided a greater variety of external reporting for review. Overall, our review indicated that customers have struggled with maintenance challenges.

As of November 2022, a Chinese source stated that there were 200 CH-4 drones of unspecified variants sold on the international market.[81] Production and sales seem to be proceeding steadily despite potential issues operating and sustaining the system.[82] The extent and structure of after-

---

[80] The PLA employs the CH-4B variant, which is a kinetic payload variant. Some of China's exports are the CH-4A variant, which is a nonkinetic intelligence, surveillance, and reconnaissance platform. The systems are similar, and, therefore, we refer to the CH-4 generally in this section. See International Institute for Strategic Studies, 2023, p. 239; and Stockholm International Peace Research Institute, "Trade Registers," webpage, undated.

[81] Liu Xuanzun, "Over 200 Chinese CH-4 Drones Sold on International Market; New Facility to Expand Production Capacity: Manufacturer," *Global Times*, November 2, 2022.

[82] Jr Ng, "China's CASC Wins Follow-On CH-4 UAV Export Order," Asian Military Review, September 16, 2022; Military Leak, "CASC to Expand Production Capacity of CH-4 Rainbow Unmanned Aerial Vehicles," November 7, 2022.

sale support provided by Chinese manufacturers is unclear, although a video released by the Iraqi Ministry of Defense of a CH-4 launch shows a Chinese support crew in the ground control station.[83] In addition, *Janes* reported in January 2020 that the China Aerospace Science and Technology Corporation—which develops the CH-4 and other UAVs—was awarded a contract from a Chinese civilian agency for fixed-wing, UAV-based surveying and mapping systems and that the contract included maintenance and aftermarket support.[84] This contract indicates that maintenance services could be included in Chinese foreign military sales, but the extent of these maintenance services is unknown.

News sources point to several maintenance-related complaints about the system, such as "sketchy service records," a "lack of servicing and maintenance documentation," and "no spare part inventory or ordering system."[85] The U.S. Inspector General reports for Combined Joint Task Force–Operation Inherent Resolve (CJTF–OIR) document numerous maintenance issues experienced by the Iraqi military when it tried to employ the CH-4. According to the April–June 2019 quarterly report:

> In addition to the reduction in Coalition ISR sorties, the ISF has struggled with the availability of its own ISR assets. CJTF-OIR reported that maintenance problems resulted in only one of Iraq's more than 10 CH-4 aircraft—Chinese unmanned aerial system (UAS) similar in design to the American MQ-9 Reaper—[being] fully mission capable.[86]

By mid-2021, no CH-4s appear to have been operational. According to the Inspector General's April–June 2021 quarterly report:

---

[83] Franz-Stefan Gady, "Revealed: Chinese Killer Drones in Iraq," *The Diplomat*, October 16, 2015.

[84] Kelvin Wong, "Update: CASC Expands Domestic UAV Testing and Production Capabilities," *Janes*, January 8, 2022.

[85] Stijn Mitzer, "Operational Failure: The CH-4B's Short-Lived Career in Jordan," *Oryx* blog, January 17, 2022; Stijn Mitzer and Joost Oliemans, "Tracking Worldwide Losses of Chinese-Made UAVs," *Oryx* blog, November 19, 2021.

[86] U.S. Department of Defense Inspector General, "Operation Inherent Resolve: Lead Inspector General Report to the United States Congress," August 2, 2019.

CJTF-OIR reported that Iraq has been unable to use its Chinese-produced CH-4 UAVs, with the last aircraft flown in September 2019. CJTF-OIR added that Iraq initially had a fleet of 20 CH-4s, but that 8 crashed. Iraqi sources informed CJTF-OIR that the reason the remaining 12 have not flown is because they are awaiting parts from China.[87]

Maintenance problems were still being reported as of 2022, which indicates the persistence of the maintenance issues and their long-term impact on operational readiness.[88] Although it is difficult to assess whether these issues are unique to foreign sales or whether the PLA endures similar issues with its own fielded CH-4 systems, there are indications that the reliability of the system could require significant maintenance support and the supply system that supports repair parts might not be robust. Analyzing the increasing Chinese exports of systems that the PLA also fields, such as Pakistan's recent acquisitions of the J-10C fighter jet, could provide additional insights regarding PLA maintenance dynamics going forward.

CH-4 customers have reported many losses for the number of drones that they have purchased. In November 2021, *Oryx* reported that Saudi Arabia had lost 12 CH-4s, Iraq had lost eight, and Algeria had lost three.[89] However, it is unclear whether the CH-4 loss rate is higher than the loss rate for U.S.-manufactured equivalent drones, such as the Predator or Reaper.

Conclusion

In our evaluation of the PLA's approach to maintenance management, we identified several emerging strengths and continued weaknesses. To close the gap in maintenance capability, the PLA has heavily invested in technology and system solutions and leveraged the benefits of MCF. Reforms in maintenance scheduling and workforce design have increased efficiencies.

---

[87] U.S. Department of Defense Inspector General, "Operation Inherent Resolve: Lead Inspector General Report to the United States Congress," August 3, 2021.

[88] U.S. Department of Defense Inspector General, "Operation Inherent Resolve: Lead Inspector General Report to the United States Congress," November 1, 2022.

[89] Mitzer and Oliemans, 2021.

However, these investments have failed to fully mitigate many of the weaknesses that exist within the workforce itself. There is a wide discrepancy in skill levels for maintainers, which creates an overreliance on senior maintainers. Additionally, organizational culture in PLA maintenance organizations tends to be risk averse, which leads to less innovation and independent decisionmaking. Innovation that does occur appears to be localized and lacks broader application across the force. Reforms of the system are still ongoing; how those reforms affect the PLA's maintenance capabilities will require further investigation over time.

CHAPTER 5

Implications and Recommendations

Understanding the PLA's approach to maintenance management is essential for assessing the PLA's ability to sustain a modernized force. As we explored in Chapters 3 and 4, the PLA is rapidly modernizing its military systems. This process will require sophisticated maintenance management practices to continue maintaining more-complex systems. In our research, we examined the PLA's historical approach to maintenance, identified critical reforms that affect maintenance practices, and highlighted key themes related to the PLA's maintenance capabilities. By describing these practices in depth, we explore the relationship between maintenance and operations and identify initial insights as to how maintenance might shape the PLA's plans to operate.

Key Findings

Our research indicated that the PLA is a maintenance force in transition and generated the following six key findings:

- The PLA has historically viewed maintenance as separate from its logistics functions. As of 2023, the JLSF reforms have had little applicability to maintenance functions.
- Primary drivers of logistics reforms include assessments of logistics failures experienced in previous conflicts, observations of technological advances in Western militaries, and desires to realize greater efficiencies within the logistics system.

- The PLA recognizes the need for a skilled maintainer force to meet the requirements presented by rapidly modernizing and sophisticated systems.
- One of the critical weaknesses in the PLA's maintenance system is the lack of a professionalized maintenance force. The PLA has made some progress in enhancing the skill level of the force by making changes to NCO training and imbuing NCOs with more responsibility. However, it is difficult to overcome the large differentiation in skill level between junior and skilled maintainers, particularly when the organizational culture does not prioritize innovation and knowledge-sharing.
- The PLA has prioritized improving its self-identified maintenance weaknesses by instituting process reforms and compensating for low skill levels by developing technological solutions. These solutions mitigate some of the gaps in PLA maintenance proficiency; process reforms are the most promising. However, the fast-paced evolution of weapon system technology might outpace some technology-based compensations.
- Initial analysis of weapon systems that are fielded to the PLA and exported abroad showed maintenance issues affecting customers. These issues include unavailable repair parts because of insufficient supply chains, with the parts shortage affecting system availability. Although it is unclear whether the PLA faces these same challenges at home, analyzing the increasing Chinese exports of the systems that the PLA fields could provide additional insights regarding maintenance going forward.

Implications

The following three implications emerged from our research:

- Poor maintenance is an indicator of technical skill deficiencies. Within the PLA, there are indications that operators and junior-level maintainers do not have the knowledge, skills, and abilities to perform some routine maintenance functions; senior maintainers shoulder a heavier maintenance burden. This skill deficiency could impede operators' ability to use systems to their full capability and is of particular interest as the PLA continues to modernize its weapon systems.

- Maintenance shapes a military's pattern of adaptation.[1] When maintenance proficiency is low, an organization can either compensate for deficiencies through operational design, or it can attempt to improve its capabilities (see Table 5.1 for a summary of the PLA's approach). Within the PLA, the system is not always responsive when addressing the needs of maintainers, as demonstrated by such deficiencies as the lack of appropriate maintenance materials and manuals that do not adequately address maintenance requirements. To overcome these deficiencies, maintainers compensate through individual efforts. However, these individual solutions do not appear to feed into systemic improvement across the organization.
- Maintenance practices can also be an indicator of how a competitor might plan to operate and give insight into their plans or assumptions regarding the duration of operations. These plans are not known for the PLA, but the stovepiped nature of the PLA's maintenance practices indicates that cross-unit interoperability could be constrained. Stovepiping could slow the PLA's tempo in operations, particularly in sustaining high-tempo operations in a stressing contingency. However, some of these constraints could be overcome by the PLA's ability to leverage the civilian sector, which would provide more flexibility.

Future Areas of Research

Exploring the PLA's historical approach to maintenance sheds light on how the PLA conceptualizes the importance of maintenance as a factor of operational success, as well as where it assesses potential weaknesses within its maintenance systems and practices that warrant improvement or compensation. With this research, we aimed to describe the maintenance practices of the PLA, and more specifically the PLAA and PLAAF, and the potential relationship between these practices and operations. The relative dearth of

---

[1] Powell, 2019, p. 30.

TABLE 5.1

Summary of PLA Approaches to Address Maintenance Deficiencies

Deficiency	Root Causes	Approach to Address Deficiency
Wide variation in skill level from junior to senior maintainers	• Dependence on senior maintainers • Lack of professional pride in maintenance profession • Textbooks and manuals that do not reflect the complexity of the maintenance required on modern weapon systems • Limited NCO reforms	• Compensation: Addressed through insufficient technology solutions • Improvement: Attempts to professionalize the NCO force and reform maintenance manuals through NCO instructor exchanges with operating units
Inefficient maintenance processes	• Lack of professional pride resulting in lack of attention to detail • Prevalence of corruption in maintenance	• Improvement: Routinization of maintenance scheduling and quality assurance reforms that increased organizational efficiency
Lack of innovation	• Risk aversion in organizational culture • Preference to adhere to hierarchical structures • Limited knowledge, skills, and abilities of maintainer force	• Compensation: Top-down initiatives that lead to inefficiencies because of bureaucracy
Poor knowledge-sharing	• Strict adherence to hierarchies that affects sharing among individuals and units • Personnel remaining within one unit	• Compensation: Unit-specific processes and work-arounds that are not broadly implemented across the force

research on this topic provides many opportunities for further research on related issues. We lay out six options.

First, most of our research focused on maintenance activities occurring in training, professional military education, and operational support. Further study could explore maintenance support to exercises. What insights might be gleaned from different types and levels of exercises, including large-scale exercises in China; branded exercises, such as "Golden Helmet"; and other service-specific exercises or competitions, or even exercise deployments to Russia?

Second, because of the emphasis on certain services in Chinese sources, our research focused on the PLAA and PLAAF, but we did identify some similar maintenance issues facing the PLAN. For example, a shortage of highly skilled troops has impeded the availability of warships in operational testing and limits naval assets from being used to their full capability.[2] A report also indicated that the low technical proficiency of key personnel, such as maintainers, will require a rebalancing of training and additional recruitment in the PLAN.[3] Further exploration of maintenance issues that affect the PLAN might identify similarities as well as unique challenges that the PLAN faces compared with the PLAA and PLAAF. The PLAN provides an interesting case for further study because of its unique ties to civilian industry.[4] Additionally, there is an opportunity to examine maintenance in other PLAA and PLAAF branches or units that are deemed less prestigious than combined arms brigades and combat aviation, such as PLAAF surface-to-air missile forces.[5] Understanding how the PLAAF conducts maintenance on these types of systems compared with aircraft could illuminate whether maintenance approaches and technical proficiency differ across systems and platforms. Finally, maintenance practices in the PLARF and PLA Strategic Support Force remain hard targets because of the limited

---

[2] Kristin Huang, "Chinese Military Short of Troops Trained in Hi-Tech Operations, PLA Daily Reveals in Rare Show of Candour," *South China Morning Post*, January 2, 2023.

[3] "Report: PLA Navy Runs into Crewing Difficulties for Growing Fleet," *Maritime Executive*, January 3, 2023.

[4] For example, in 2009, the PLAN established its first integrated civil-military vessel equipment center. See Puska, 2010.

[5] For more information on PLAAF surface-to-air missiles, see Bonny Lin and Cristina L. Garafola, *Training the People's Liberation Army Air Force Surface-to-Air Missile (SAM) Forces,* RAND Corporation, RR-1414-AF, 2016.

information on them that is available to the public, but both would also be valuable to study.

Third, as our exploration of the CH-4 UAV exports showed, international customers have experienced issues with repair parts supply chains that have affected system availability. It is unclear whether China experiences similar issues. Through our research, we were also able to glean some insights into the extent and structure of after-sale support. However, our ability to investigate other cases was constrained by a lack of common systems that China both exports overseas and fields at home within the PLA and by the difficulty of matching recent Chinese articles about maintenance practices to specific systems. As China continues to grow the volume, value, and sophistication of the weapon systems that it exports abroad, studying maintenance-related agreements for deliveries of such systems as the J-10C fighter aircraft—which was delivered to Pakistan in 2022 and is fielded by the PLA—might help further illuminate maintenance practices and challenges that are relevant for Chinese maintenance forces.

Fourth, the JLSF reforms have been a focus for researchers seeking to assess the PLA's logistics capabilities. During our research, however, we did not find indications that these reforms have significantly affected maintenance functions. Areas for further study might include identifying how maintenance interacts with other functions of logistics, such as supply and transportation, that have been affected by restructuring under these reforms. For instance, how might the health of the defense industrial base affect maintenance operations?

Fifth, a natural next step for the research is exploration of how maintenance functions directly affect PLA operations. How does maintenance acumen affect the PLA's ability to conduct certain types of operations? More broadly, maintenance performance is linked to an organization's technical skill level and indicates its ability to adapt. How do we assess that the PLA's technical skill level could affect operational capabilities?

Sixth, insights from the first five topics could support analysis that identifies and assesses the PLA's likely maintenance requirements and challenges for a large-scale contingency, such as a Taiwan campaign.[6] This avenue of

---

[6] For some initial exploration of China's logistics capability for a Taiwan campaign, see Jacob Maywald, Benjamin Hazen, Edward Salo, and Michael Hugos, "Logistics Inter-

research would also benefit from synthesizing PLA observations and lessons learned from Russia's invasion of Ukraine.[7] How a maintenance system functions can serve as an indicator of a competitor's expectations of conflict and its duration. Do the PLA's maintenance capabilities hinder or enhance its ability to project power and achieve strategic goals? All six areas are worthy of further analysis.

---

diction for Taiwan Unification Campaigns," *War on the Rocks*, August 21, 2023.

[7] For an early example of PLA analysis on this topic, see Wang Fengcai [王凤才], "The Logistical Problems Russia Exposed in the Russia-Ukraine Conflict" [俄乌冲突中俄军战场后勤保障暴露的问题], *World Military Review* [外国军事学术], No. 5, 2022.

Other Logistics Success Factors

In this appendix, we detail five logistics success factors that we identified as part of the research summarized in Chapter 3. Table A.1 lists each factor, provides a detailed description of it, and offers examples of success or failure related to the factor.

We opted not to further analyze PLA maintenance processes and practices for these five factors because they have less direct relevance to the maintenance subfunction of logistics, there was not much information readily available on these factors in the literature on PLA maintenance, or both. Further research examining PLA logistics processes and practices could potentially incorporate these success factors as well.

Integration of Logistics Personnel into Operational Units

Closely related to balanced emphasis of support and combat elements is the extent to which logistics personnel are integrated into operational units. Integration, in this sense, could take various forms, including the colocation of logistics personnel with operational units, the use of military exercises that include both operations and support elements working together in realistic scenarios, and the extent to which logistics plans are integrated and coordinated with strategic ones. Historical case studies provide clear illustrations of the operational implications of this concept. For example, Kane describes how Admiral Richmond Kelly Turner, who commanded the amphibious force during the Pacific War, prioritized ensuring that sailors and logisticians trained together and on the same ships to which they would

TABLE A.1

Five Factors Present in Successful Logistics Organizational Cultures

Factor in Success	Description of Factor and Results	Examples of Operational or Strategic Success or Failure
Integration of logistics personnel into operational units	• Colocation of logistics personnel with operational units • Use of military exercises that include both operations and support elements working together in realistic scenarios • Integration and coordination of logistics plans with strategic plans	• United States in the Pacific War, 1941–1945 • World War I, Mobile Ordnance Repair Shops, 1917–1918
Unified chain of command	• Unified command structure with a clear chain of command • Clear lines of communication • Span of control	• U.S. military information operations, 2010s
Appropriate balance between competing priorities in a constrained environment	• Appropriately balanced cost, quality, and timeliness factors related to maintenance and supply • No prioritization of cost, quality, or timeliness factors at the expense of each other	• Flight operations on U.S. Navy carriers, 1980s
Low prevalence of corruption	• Lack of corruption that diverts scarce resources and has negative effects on operational performance	• Russian invasion of Ukraine, 2022
Balanced emphasis on support and combat elements	• Sufficient support to combat soldier ratio • Protection provided to support units	• Russian invasion of Ukraine, 2022 • United States in the Pacific War, 1941–1945

be assigned in combat.[1] He emphasizes the value this coordinated training provided: "Marine logisticians and sailors had a great deal of difficulty working together during rehearsal exercises off San Clemente Island, but

---

[1] Kane, 2001.

they had managed to solve their problems almost completely by the invasion itself."[2] Another facet of logistical integration is the mobility of combat support relative to operations. Hirrel, for example, describes how, during World War I, Mobile Ordnance Repair Shops (moveable maintenance shops that operated out of trucks and tents) emerged as complements to fixed-base shops that were responsible for extensive overhauls.[3] Mobile maintenance allowed U.S. forces to respond more flexibly to in-theater maintenance needs and subsequently endured as a maintenance practice throughout the 20th century.

Previous studies have explored the idea that combat support should be just as mobile as operations. In a 2021 RAND report on distributed operations that require quick changes in the location of the ground support for air operations, the researchers found that such operations place unique demands on combat support command and control, particularly in terms of the requirement to coordinate activities and needs between the operational and combat support communities.[4] Snyder et al. found that the "physical limits of what resources are available and how rapidly they can move will constrain these operations . . . . If combat support is not sufficiently coordinated with combat operations, plans might not be supportable and the mission could fail." In a 2015 RAND report on contingency planning, the researchers found that combatant commanders and their component commands often lack information about the availability of global combat support resources, which results in serious operational risks.[5] They attributed this gap to a disconnect between operations and combat support communities and to the lack of a coordinated approach (including "doc-

---

[2] Kane, 2001.

[3] Leo P. Hirrel, "World War I and the Emergence of Modern Army Sustainment," in Keith R. Beurskens, ed., *The Long Haul: Historical Case Studies of Sustainment in Large-Scale Combat Operations*, Army University Press, 2018.

[4] Don Snyder, Kristin F. Lynch, Colby P. Steiner, John G. Drew, Myron Hura, Miriam E. Marlier, and Theo Milonopoulos, *Command and Control of U.S. Air Force Combat Support in a High-End Fight*, RAND Corporation, RR-A316-1, 2021.

[5] Robert S. Tripp, John G. Drew, and Kristin F. Lynch, *A Conceptual Framework for More Effectively Integrating Combat Support Capabilities and Constraints into Contingency Planning and Execution*, RAND Corporation, RR-1025-AF, 2015.

trine, processes, analytic tools, training regimen, and organizations") for "systematically includ[ing] CS [combat support] resource capabilities and constraints within the contingency planning process."[6] To be effective, combat support must be integrated with operational units both spatially and organizationally.

Unified Chain of Command

As mentioned previously, literature on the performance of centralized versus decentralized organizations is mixed. However, in the military context, and particularly regarding logistics, a unified command structure with a clear chain of command might be more conducive to meeting operational demands. Min, for example, argues against the applicability of organizational theories that favor decentralized management structures in complex environments to U.S. military information operations, citing significant differences between military and civilian organizations.[7] These differences are related, in part, to the high degree of uncertainty and danger inherent in military operations. As Dillard and Nissen argue, when "project risk or quality is paramount, formalized procedures, vertical information flows, and centralized decision-making typical of bureaucratic organizational forms can be seen as superior."[8] In the logistics domain, the 2021 RAND report on distributed operations found that effective combat support in such operations requires a unified command and control mechanism at the operational level.[9] Without this mechanism, the speed of combat support is "slowed down by the need for the coordinated actions of many actors, both within and outside the theater, including service, joint, and agency actors."[10]

---

[6] Tripp, Drew, and Lynch, 2015, pp. 1–2.

[7] Ryan B. Min, *Limits of Decentralization: Streamlining the Dispersed Parts of the Information Whole*, thesis, U.S. Army Command and General Staff College, 2018.

[8] John Dillard and Mark E. Nissen, *Determining the Best Loci of Knowledge, Responsibilities and Decision Rights in Major Acquisition Organizations*, Naval Postgraduate School, June 30, 2005, p. 34.

[9] Snyder et al., 2021.

[10] Snyder et al., 2021, p. viii.

Span of control, or the number of subordinates a supervisor can effectively manage, is related to the concept of chain of command. Although studies of military organizations do not specify an optimal supervisor-to-subordinate ratio, they do provide some insight into potential indicators of the appropriate span of control. For example, Benton notes that span of control is closely tied to *departmentation,* or the "groupings of both people and tasks" based on "logical divisions of work."[11] Specifically, departmentation has the potential to "simplify managerial tasks and maintain control by grouping employees within well-defined areas," so that as this type of division of work increases, so too does the potential span of control of a department manager.[12] Elaborating on this concept, Pierce outlines the organizational factors that influence span of control, including similarity and complexity of functions, geographic distance between manager and subordinates, and the rate at which the external environment changes.[13] Comparing these factors with the actual organization of a logistics entity—whether at the unit level or more broadly across the military organization—can therefore provide some indication of the entity's potential effectiveness in achieving its objectives.

Appropriate Balance Between Competing Priorities in a Constrained Environment

Resources—money, time, and people—will always be limited. An organization's performance is therefore closely tied to its ability to balance competing priorities in a constrained environment. This principle is particularly applicable to military logistics and sustainment. Trade-offs related to the maintainability and reliability of a system during its design offer one example of potential tensions between competing priorities. As Alexander observes, "one of the tradeoffs that continually beset program managers [is] with fixed development budgets, they have to choose where to spend their

---

[11] Lewis R. Benton, *Supervision and Management,* McGraw-Hill Book Company, 1972.

[12] Benton, 1972.

[13] William G. Pierce, *Span of Control and the Operational Commander: Is It More Than Just a Number?* thesis, U.S. Army Command and General Staff College, 1991.

money . . . . They must allocate available resources into improving reliability, lowering production cost, or raising performance."[14] Clarke elaborates on this idea: "While designing for [maintainability and reliability] may cost very little, it might also cost a great deal. The methodology of life cycle cost is concerned with the trade-off of the impact of designing for increased reliability and maintainability on operating and support costs, against any additional [research and development] and production cost that might be incurred."[15] Importantly, reliability (which maintainers seek to achieve) is inversely related to performance (which operators prioritize). As Alexander notes, "performance is a price that can be paid for higher reliability. . . . Technology can loosen constraints, but it does not eliminate the need for assigning priorities and considering tradeoffs."[16]

In the military context, in which complex systems are operated in highly uncertain and dangerous environments, efficiency must be weighed against other, competing priorities, such as safety. For example, Sagan describes a common characteristic of high-reliability organizations as the ability to "enjoy abundant resources so that short term efficiencies can be neglected in favour of reliable operation."[17] In their study of civilian maintenance engineers, Pettersen and Aase observe a similar phenomenon:

> Institutionalisation of " . . . the value of slowing things down to enhance reliability and address all safety concerns" guided the technicians' actions in critical situations of sensemaking and gave legitimacy to the creation of extended performance spaces (e.g., timeframes) for making sure that the airplane was technically airworthy.[18]

---

[14] Arthur J. Alexander, *The Costs and Benefits or Reliability in Military Equipment*, RAND Corporation, P-7515, 1988, p. 11.

[15] John D. Clarke, *Lifecycle Cost: An Examination of Its Application in the United States, and Its Potential for Use in the Australian Defense Forces*, thesis, Naval Postgraduate School, 1990, p. 13.

[16] Alexander, 1988, p. 11.

[17] Scott D. Sagan, *The Limits of Safety: Organizations, Accidents, and Nuclear Weapons*, Princeton University Press, 1993, as cited in Pettersen and Aase, 2008.

[18] Pettersen and Aase, 2008, p. 515.

Redundancy—which enables the execution of a task if the primary unit designated to complete the task fails—is one mechanism through which efficiency might be sacrificed in favor of safety. Rochlin, La Porte, and Roberts describe the significance of supply redundancy for Navy aircraft carrier flight operations and the careful trade-offs it requires:

> The ship must carry as many aircraft and spares as possible to keep its power projection and defensive capability at an effective level in the face of maintenance requirements and possible operational or combat losses. . . . Here is a clear case of a trade-off between operational and safety reliability that must be made much closer to the edge of the envelope than would be the case for other kinds of organizations.[19]

Low Prevalence of Corruption

Corruption diverts scarce resources, which can negatively affect operational performance. For example, analysts have pointed to systemic corruption in Russia's defense procurement sector, which has resulted in soldiers receiving inadequate equipment and supplies, as a defining factor in Russia's lackluster operational performance in Ukraine.[20] Tagarev outlines the various pathways through which corruption degrades military effectiveness.[21] These pathways include negatively affecting the caliber of armed forces through promotions or appointments obtained by bribery versus competency or capability, negatively affecting discipline by undermining the trust in and respect for the chain of command, and decreasing the value of defense expenditures by "diverting resources from the appropriately priced generation of defense capabilities."[22]

---

[19] Rochlin, La Porte, and Roberts, 1987, p. 85.

[20] Polina Beliakova, "Russian Military's Corruption Quagmire" *Politico*, March 8, 2022.

[21] Todor Tagarev, "Enabling Factors and Effects of Corruption in the Defense Sector," *Connections*, Vol. 9, No. 3, Summer 2010. See also Thomas H. Au, "Combating Military Corruption in China," *Southern Illinois University Law Journal,* Vol. 3, No. 2, Winter 2019.

[22] Tagarev, 2010, p. 78.

Questions related to the organizational-level cultural factors associated with both corruption and military effectiveness are difficult to answer. For example, what are the characteristics of organizations with higher levels of corruption, and do these characteristics necessarily reduce military effectiveness? Hofstede's framework for measuring culture provides a starting point for understanding these relationships.[23] The framework includes four dimensions of culture: *power distance*, or the extent to which an unequal power distribution is accepted in the society; *individualism*, or the level of emphasis on an individual's responsibility to themselves versus their social network; *masculinity*, or the extent to which a society prioritizes such things as achievement and assertiveness; and *uncertainty avoidance*, which captures a society's level of risk aversion. Applying this framework to data on commercial firms in Vietnam, Nguyen et al. found that a chief executive officer's cultural background influences the effect of corruption on a firm's financial performance.[24] Specifically, they found that the negative impact of corruption on financial performance is more severe when the chief executive officer comes from a culture that is risk averse and individualistic. Tying these cultural characteristics to operational performance, Fowler uses data on battles spanning military engagements from 1600 to 1990 to test the relationship between Hofstede's cultural dimensions and battle effectiveness.[25] He finds that risk-averse and individualistic cultures (the same characteristics that Nguyen et al. found were positively associated with more-severe effects of corruption on financial performance) to be negatively associated with battle effectiveness. In other words, organizations with higher levels of corruption might have cultural characteristics that are detrimental to operational performance in ways that go beyond the immediate effects of corruption.

---

[23] Geert Hofstede, *Culture's Consequences: International Differences in Work-Related Values*, Sage Publications, 1984.

[24] Hieu Thanh Nguyen, Kieu Trang Vu, Loan Quynh Thi Nguyen and Hiep Ngoc Luu, "CEO Culture, Corruption, and Firm Performance," *Applied Economics Letters*, Vol. 29, No. 7, April 2022.

[25] Eric S. Fowler, *Culture and Military Effectiveness: How Societal Traits Influence Battle Outcomes*, thesis, Old Dominion University, Spring 2016.

Balanced Emphasis on Support and Combat Elements

In any military operation, support and combat elements are reliant on each other. Support elements, including logistics and its subfunctions, cannot survive without adequate protection from warfighters. Warfighters, in turn, cannot carry out their mission without the critical equipment, supplies, and infrastructure generated by support elements. The successful execution of any military mission is therefore contingent on an organization's ability to appropriately balance its combat and support elements.[26] The *tooth-to-tail ratio*, or the ratio of combat to noncombat personnel, is one measure of the extent to which these elements are balanced.[27] Although the literature does not identify an optimal tooth-to-tail ratio, Shrader does note that "modern, complex, mechanized, and technologically sophisticated armies, operating worldwide in every conceivable climate and terrain . . . require that a significant portion of the total force [be] dedicated to providing the required logistical support to the few who actually do the fighting."[28]

Another factor related to the coequal emphasis of support and combat elements is the level of protection offered to support units in combat. For example, studies have linked the effectiveness of military operations to the ability to maintain stable *ground lines of communication*, which are routes along which supplies and forces are moved in support of operations.[29] If ground lines of communication are jeopardized, unsupported forces lose their ability to sustain operations.[30] Further emphasizing the importance of protecting support elements, Kane describes how, during the Guadalcanal

---

[26] Allan R. Millet, Williamson Murray, and Kenneth H. Watman, "The Effectiveness of Military Organizations," in Allan R. Millet and Williamson Murray, eds., *Military Effectiveness, Volume 1, The First World War*, Cambridge University Press, 1988.

[27] John J. McGrath, *The Other End of the Spear: The Tooth-to-Tail Ratio (T3R) in Modern Military Operations*, Combat Studies Institute Press, 2007.

[28] Charles R. Shrader, *U.S. Military Logistics, 1607–1991: A Research Guide*, Greenwood, 1992.

[29] David J. Kolleda, *Force Protection Through Security of the Ground Lines of Communication (GLOC)*, U.S. Army Command and General Staff College, 1995.

[30] Kolleda, 1995.

campaign in the Pacific theater of World War II, U.S. troops were stranded with half of their food and such critical supplies as artillery, radios, radar gear, and heavy weapons because of the Navy's decision to provide only two days of air cover for support units; this situation ultimately impeded and prolonged the brutal campaign.[31] In a more contemporary example, military analysts commenting on Russia's February 2022 invasion of Ukraine have noted that Russia failed to adequately protect trucks along supply lines. As a result, those units fell victim to counterattacks by Ukrainian forces that, in part, undermined plans for a quick takeover of Kyiv.[32]

---

[31] Kane, 2001.

[32] Berkowitz and Galocha, 2022.

People's Liberation Army Logistics and Maintenance Terms of Reference

In this appendix, we list key organizational and operational terms relevant for PLA logistics and maintenance practices. Each term is first listed in English, then in Chinese. We include sources to provide examples of use. For terms that are defined or explained in PLA dictionaries or encyclopedias, we include translations of those definitions or descriptions. Unless otherwise stated, English versions of terms from the PLA dictionary and PLAAF encyclopedia are original to those sources.

Key People's Liberation Army Logistics Organizations

The following are key PLA logistics organizations:

- Joint Logistics Support Force (JLSF) (联勤保障部队)[1]
- CMC Logistics Support Department (中央军委后勤保障部)[2]

---

[1] This is the most common term used for the JLSF; longer versions add CMC or PLA in front of JLSF. For an example of use, see "The Ministry of National Defense Introduced the Construction and Development of the Joint Logistic Support Force Since Its Reform and Reconstruction" [国防部介绍联勤保障部队改革重塑以来建设发展情况], Ministry of National Defense Network [国防部网], September 30, 2021.

[2] Sun Xingwei [孙兴维] and Zhang Peng [张鹏], "The Logistic Support Department of the Central Military Commission Issued the 'Military Personnel Health Examina-

- JLSF Headquarters (联勤保障部队司令部), more commonly referred to as the Wuhan Base (武汉联勤保障基地)[3]
- Joint Logistics Support Center(s) (联勤保障中心)[4]

General Logistics and Maintenance Terms from the People's Liberation Army Military Terminology Dictionary (中国人民解放军军语) (2011)

Logistics Terms

The following are relevant logistics terms from the PLA dictionary:[5]

- *Precision logistics support* (后勤精确保障): "Using modern information technology and other sub-segments of high technology, organize the implementation of logistics support that is timely, at the appropriate location, of the appropriate quality, and is applicable [of the appropriate use] for the force."[6]

---

tion Measures,'" [中央军委后勤保障部印发"军队人员健康体检办法"], Xinhua [新华社], April 27, 2022.

[3] "In 1985, the Wuhan Military Region Was Dismantled, and the Henan Province and Hubei Province Military Regions Within Its Jurisdiction; How to Transfer and Adjust?" [1985年，武汉军区撤编，辖内河南省、湖北省军区，如何转隶调整？], Cyclical Rate History [周期率历史], January 27, 2023.

[4] See, for example, Lu Ke [卢科] and Sun Xingwei [孙兴维], "The Joint Logistic Support Force Innovates with Three Mechanisms to Comprehensively Upgrade the Quality of General Engineering Construction" [联勤保障部队创新3项机制 全面提升通用工程建设质量], *PLA Daily* [解放军报], June 4, 2022; Sun Xingwei and Zhang Liang, "Chinese Military Organizes Special Production Training for Peacekeepers for First Time," *China Military Online*, June 21, 2021.

[5] We exclude some terms that do not seem to be in recent use, such as contingency logistics support brigade (后勤应急保障旅). See Academy of Military Science All-Military Military Terminology Committee, 2011, p. 476. We also exclude terms that have outdated definitions following the 2015–2016 PLA reforms, such as logistics organ ([后勤机关), which references the now disbanded GLD. See Academy of Military Science All-Military Military Terminology Committee, 2011, p. 327.

[6] Academy of Military Science All-Military Military Terminology Committee, 2011, p. 485. In English, this dictionary is also referred to as the *PLA Dictionary* or *Junyu* (军语).

- *Logistics support echelon* (后勤保障环节): "In the military logistics support system, a first-level logistics organization with organizational planning and comprehensive support functions, such as a theater joint logistics department and its subordinate logistics forces, a division logistics department and subordinate logistics forces, etc."[7]
- *Joint logistics branch* (联勤分部): "An organization dispatched by the Joint Logistics Department of a military region (theater command) to undertake the tasks of regional support, logistics base construction, and management. It has a headquarters, a political department, and a number of business divisions (处). It has jurisdiction over hospitals, warehouses, transportation, and other service forces and elements (分队). It is the basic force of joint logistics support."[8]
- *Logistics detachment* (后勤分队): "A service support unit responsible for logistical support tasks; such as a motorized battalion, pipeline team, service company, etc."[9]
- *Logistics division* (后勤处): "The leading agency in charge of logistical work in regiment-level forces or units as established by the force. Under the leadership of that force or the head of that unit. Responsible for leading and managing various logistics services, and organizing and implementing logistics support and other work."[10]
- *Logistics officer* (后勤军官): "Officers engaged in logistical work; includes military officers engaged in logistics command, finance, quartermaster, supplies, fuel, transportation, sanitation, barracks, and other work."[11]

---

[7] Academy of Military Science All-Military Military Terminology Committee, 2011, p. 476.

[8] Academy of Military Science All-Military Military Terminology Committee, 2011, p. 476.

[9] Academy of Military Science All-Military Military Terminology Committee, 2011, p. 477.

[10] Academy of Military Science All-Military Military Terminology Committee, 2011, pp. 327–328.

[11] Academy of Military Science All-Military Military Terminology Committee, 2011, p. 340.

Equipment and Maintenance Terms

The following are relevant equipment and maintenance terms from the PLA dictionary:

- *Management of communications on duty and maintenance* (通信值勤维护管理): "To carry out the management of communications duty personnel and their on-duty and maintenance work. Including combat readiness duty and command dispatch management, communications duty management, management of machine line equipment and maintenance, network operations management, management of communication stations, and security management, etc."[12]

- *Communications inspection and maintenance* (通信检修): "Activities to carry out inspections and maintenance of communications equipment, materials, lines and communication networks, systems, etc."[13]

- *Communications equipment maintenance* (通信装备维护): "Maintenance carried out in order to keep communications equipment in good condition. Mainly [involves] wiping clean the equipment, inspecting, maintaining, testing, identifying and checking, replenishing consumed [items], etc."[14]

- *Equipment maintenance and management expenses* (装备维修管理费): "Special funds used by the military to maintain, keep in good repair, and manage weapons and equipment, to purchase maintenance equipment and materials, to organize specialized training and related business activities, etc."[15]

- *Integrated military-civilian equipment maintenance support* (军民一体化装备维修保障): "In order to improve the equipment maintenance support capability and support effectiveness, using the military sup-

---

[12] Academy of Military Science All-Military Military Terminology Committee, 2011, p. 249.

[13] Academy of Military Science All-Military Military Terminology Committee, 2011, p. 249.

[14] Academy of Military Science All-Military Military Terminology Committee, 2011, p. 250.

[15] Academy of Military Science All-Military Military Terminology Committee, 2011, p. 490.

port forces as the main body, coordinate the support mode of the military and local equipment maintenance support programs and plans, resource allocation, and force employment."[16]

- *Equipment maintenance support* (装备维修保障): "To carry out maintenance and repair activities in order to maintain and restore equipment to good technical condition or to improve equipment performance."[17]

- *Equipment maintenance support system* (装备维修保障体制): "The organizational system of systems and corresponding system for equipment maintenance support. Includes the machine settings, functional divisions, interrelationships, etc. of equipment maintenance support."[18]

- *Equipment maintenance support plan* (装备维修保障计划): "A plan formulated for organizing the implementation of equipment maintenance support. Includes medium- and long-term plans, annual plans, and special plans, etc. for equipment maintenance."[19]

- *Equipment maintenance support resources* (装备维修保障资源): "A collective term that can be used for equipment maintenance support manpower, material resources, financial resources, information, etc."[20]

- *Equipment maintenance support facilities* (装备维修保障设施): "A collective term for permanent or semi-permanent buildings, sites, and ancillary equipment required for equipment maintenance support. Includes maintenance sites, maintenance workshops, maintenance equipment warehouses, etc."[21]

---

[16] Academy of Military Science All-Military Military Terminology Committee, 2011, p. 546.

[17] Academy of Military Science All-Military Military Terminology Committee, 2011, pp. 548–549.

[18] Academy of Military Science All-Military Military Terminology Committee, 2011, p. 549.

[19] Academy of Military Science All-Military Military Terminology Committee, 2011, p. 549.

[20] Academy of Military Science All-Military Military Terminology Committee, 2011, p. 549.

[21] Academy of Military Science All-Military Military Terminology Committee, 2011, p. 549.

- *Equipment maintenance support devices* (装备维修保障设备): "A collective term for machine tools, instruments, meters, etc. used for equipment maintenance support."[22]
- *Equipment maintenance support capabilities statistics* (装备维修保障实力统计): "Statistics on relevant data of actual and directly usable strengths of equipment maintenance support for a certain period of time. Includes statistics on the quantity and quality of currently available equipment maintenance personnel and their technical level, facilities and equipment, supplies and materials, etc."[23]
- *Equipment maintenance* (装备维修): "Activities carried out to maintain and repair equipment in order to maintain and restore the stipulated technical condition or to improve equipment performance. According to the characteristics and purpose of the maintenance, it can be divided into preventive maintenance, restorative maintenance, and improvement maintenance; according to the maintenance structure and level, it can be divided into grassroots-level maintenance, intermediate-level maintenance, and base-level maintenance."[24]
- *Equipment officer* (装备军官): "Officers engaged in equipment work. Includes officers engaged in equipment development, support, management, and command work."[25]

The following term is also related to logistics and maintenance more broadly:

- *Strategic power projection* (战略投送), sometimes also translated as strategic delivery: "In order to achieve a certain strategic purpose, the action of comprehensively using various transport forces to put groups

---

[22] Academy of Military Science All-Military Military Terminology Committee, 2011, p. 549.

[23] Academy of Military Science All-Military Military Terminology Committee, 2011, p. 549.

[24] Academy of Military Science All-Military Military Terminology Committee, 2011, pp. 549–550.

[25] Academy of Military Science All-Military Military Terminology Committee, 2011, p. 340.

of armed forces into combat or crisis areas. Usually organized by the supreme command (统帅部)."[26]

Aviation-Specific Maintenance Terms from the China Air Force Encyclopedia (中国空军百科全书) (2005)

The following are relevant, aviation-specific maintenance terms from the China Air Force Encyclopedia:

- *Aircraft maintenance* (飞机维修): "maintenance and repair activities carried out in order to maintain and restore the stipulated technical condition of aircraft and helicopters."[27]
- *Aircraft maintenance specialty* (飞机维修专业): "a professional technical category for dividing up maintenance work based on the structural and functional characteristics of aircraft, helicopters, and their systems equipment and parts."[28]
- *Aviation maintenance management* (航空维修管理): "carrying out the planning, organization, control, and coordination activities of aviation equipment maintenance."[29]

Other Organizations and Terms

- Service support battalion (保障营)[30]

---

[26] Academy of Military Science All-Military Military Terminology Committee, 2011, p. 58.

[27] China Air Force Encyclopedia Editorial Committee [中国空军百科全书编审委员会], *China Air Force Encyclopedia (Vol. 1)* [中国空军百科全书 (上卷)], Aviation Industry Press [航空工业出版社], 2005, p. 551.

[28] China Air Force Encyclopedia Editorial Committee, 2005, p. 553.

[29] China Air Force Encyclopedia Editorial Committee, 2005, p. 554.

[30] Jia, Yang, and Xiang, 2018.

- Repair platoon (修理排)[31]
- Equipment maintenance office (装备维修科)[32]
- Materiel suppliers (物资供应商)[33]
- Maintenance squadron battalion-leader grade (机务大队)[34]
- Maintenance squadron company-leader grade (机务中队)[35]
- Ground crews (地勤人员)[36]
- Group level repair shops (修理厂)[37]

---

[31] See, for example, Ruan Yongjun [阮拥军] and Wang Lihong [王利洪], "On the Application of 3D Printing Technologies in Equipment Maintenance Support in Our Military's United Nations Peacekeeping Operations" [试述3D打印技术在我军联合国维和行动装备维修保障中的运用], *National Defense* [国防], No. 8, 2019, p. 39.

[32] See, for example, Chen, Feng, and Zhou, 2022; and Lei, Wen, and Li, 2021.

[33] See, for example, Yang Xueming [杨学铭] and Xun Ye [荀烨], "Study on Theater Ground Force Materiel Distribution and Support Network Mode Under the New System" [新体制下战区陆军物资配送保障网络模式研究], *Logistics Technology* [物流技术], No. 6, 2018, p. 49.

[34] See, for example, Li et al., 2021. See also Allen and Garafola, 2021, p. 127.

[35] See, for example, "Why Does Our Military's J-10 Have Two Production Batch Numbers and Parts That Come from Three Fighters?" [我军这架歼10为何有两个生产批号零件来自3架战机], *Sina Military* [新浪军事], January 11, 2019. See also Allen and Garafola, 2021, p. 127.

[36] See, for example, Xiong Chencheng [熊晨成] and Hu Ansheng [胡安胜], "Research on Coordinates-Based ABC Classification Management of Loaded Materiel" [基于坐标系的被装物资 ABC 分类管理研究], *Logistics Technology* [物流技术], No. 12, 2018, p. 144.

[37] Shi and Guo, 2020.

Abbreviations

CCP	Chinese Communist Party
CJTF–OIR	Combined Joint Task Force–Operation Inherent Resolve
CH-4	Cai Hong-4
CMC	Central Military Commission
CPVF	Chinese People's Volunteer Force
DoD	U.S. Department of Defense
EDD	Equipment Development Department
GAD	General Armament Department
GJB	*guojia junyong biaozhun*
GLD	General Logistics Department
GSD	General Staff Department
JLD	Joint Logistics Department
JLSF	Joint Logistics Support Force
LCM	life-cycle management
MCF	military-civil fusion
NATO	North Atlantic Treaty Organization
NCO	noncommissioned officer
NDU	National Defense University
PAP	People's Armed Police
PLA	People's Liberation Army
PLAA	PLA Army
PLAAF	PLA Air Force
PLAN	PLA Navy
PLARF	PLA Rocket Force
PRC	People's Republic of China
SMS	*Science of Military Strategy*
SOP	standard operating procedure

TC	theater command
TCAF	Theater Command Air Force
UAV	unmanned aerial vehicle

Bibliography

"A Perspective on China's Military Logistics Transformation: Substantial Steps in Four Areas (2)" [透视中国军事后勤大变革：四方面迈出实质性步伐(2)], *China News* [中国新闻网], January 12, 2007.

Academy of Military Science All-Military Military Terminology Committee [全军军事术语管理委员会，军事科学院], *Chinese People's Liberation Army Military Terminology* [中国人民解放军军语], Academy of Military Science Publishing House [军事科学出版社], 2011.

Alexander, Arthur J., *The Costs and Benefits or Reliability in Military Equipment*, RAND Corporation, P-7515, 1988. As of June 1, 2023: https://www.rand.org/pubs/papers/P7515.html

Allen, Kenneth W., and Cristina L. Garafola, *70 Years of the PLA Air Force*, China Aerospace Studies Institute, April 12, 2021.

Allen, Kenneth, "The Evolution of the PLA's Enlisted Force: Conscription and Recruitment (Part One)," *China Brief*, Vol. 22, No. 1, January 14, 2022.

Argote, Linda, "Organizational Learning Research: Past, Present and Future," *Management Learning*, Vol. 42, No. 4, September 2011.

Army Technical Publication, *ATP 4-33: Maintenance Operations*, Department of the Army, July 2019.

Arostegui, Joshua, and James R. Sessions, "PLA Army Logistics," *PLA Logistics and Sustainment: PLA Conference 2022*, U.S. Army War College Press, 2023.

"As the 'Gatekeeper' of Aircraft Safety, He Is Willing to Be the 'Bad Guy'" [作为机务安全"把关人"，他甘当"坏人"], *China Military Online* [中国军网], March 30, 2018.

Ashby, Mark, Caolionn O'Connell, Edward Geist, Jair Aguirre, Christian Curriden, and Jonathan Fujiwara, *Defense Acquisition in Russia and China*, RAND Corporation, RR-A113-1, 2021. As of June 1, 2023: https://www.rand.org/pubs/research_reports/RRA113-1.html

Au, Thomas H., "Combating Military Corruption in China," *Southern Illinois University Law Journal*, Vol. 3, No. 2, Winter 2019.

Baird, Lloyd, John C. Henderson, and Stephanie Watts, "Learning from Action: An Analysis of the Center for Army Lessons Learned (CALL)," *Human Resource Management*, Vol. 36, No. 4, December 1997.

Beliakova, Polina, "Russian Military's Corruption Quagmire" Politico, March 8, 2022.

Benton, Lewis R., *Supervision and Management*, McGraw-Hill Book Company, 1972.

Berkowitz, Bonnie, and Artur Galocha, "Why the Russian Military Is Bogged Down by Logistics in Ukraine," *Washington Post*, March 30, 2022.

Bierly, Paul, and Alok Chakrabarti, "Generic Knowledge Strategies in the U.S. Pharmaceutical Industry," *Strategic Management Journal*, Vol. 17, No. S2, Winter 1996.

Bjorge, Gary J., *Moving the Enemy: Operational Art in the Chinese PLA's Huai Hai Campaign*, Combat Studies Institute Press, 2004.

Blasko, Dennis J., "PLA Ground Forces Lessons Learned: Experience and Theory," in Laurie Burkitt, Andrew Scobell, and Larry M. Wortzel, eds., *The Lessons of History: The Chinese People's Liberation Army at 75*, Strategic Studies Institute, July 2003.

Blasko, Dennis J., *The Chinese Army Today: Traditions and Transformation for the 21st Century*, Routledge, 2012a.

Blasko, Dennis J., "Clarity of Intentions: People's Liberation Army Transregional Exercises to Defend China's Borders," in Roy Kamphausen, David Lai, and Travis Tanner, eds., *Learning by Doing: The PLA Trains at Home and Abroad*, Strategic Studies Institute, 2012b.

Blasko, Dennis J., "Corruption in China's Military: One of Many Problems," *War on the Rocks*, February 16, 2015.

Burk, James, "Military Culture," in Lester R. Kurtz and Jennifer E. Turpin, eds., *Encyclopedia of Violence, Peace, and Conflict. Vol. 1*, Academic Press, 1999.

Burke, Edmund J., Astrid Stuth Cevallos, Mark R. Cozad, and Timothy R. Heath, *Assessing the Training and Operational Proficiency of China's Aerospace Forces: Selections from the Inaugural Conference of the China Aerospace Studies Institute*, RAND Corporation, CF-340-AF, 2016. As of June 1, 2023: https://www.rand.org/pubs/conf_proceedings/CF340.html

Cancian, Mark F., Matthew Cancian, and Eric Heginbotham, *The First Battle of the Next War: Wargaming a Chinese Invasion of Taiwan*, Center for Strategic International Studies, January 2023.

Che Dongwei [车东伟] and Zhang Tong [张童], "A Brigade of the 72nd Group Army Focuses on Improving the Overall Level of the Equipment Maintenance Team" [第72集团军某旅注重提升装备维修队伍整体水平], *PLA Daily*, August 17, 2018.

Chen Dianhong [陈典宏], Feng Dengya [冯邓亚], and Zhou Yupeng [周宇鹏], "The Maintenance Center Has Become a 'Smart Factory,' and a Brigade of the 75th Army Group Strives to Improve the Level of Equipment Management Informatization" [维修中心变"智能工厂", 第75集团军某旅着力提升装备管理信息化水平], *PLA Daily* [解放军报], March 2, 2022.

Chen Dianhong [陈典宏] and Huang Yuanli [黄远利], "Second Sergeant Major Gan Zhanyong: Use a Pair of 'Iron Sand Palms' to Become a Tank 'Maintenance Master'" [二级军士长甘战永: 用一双"铁砂掌"练成坦克"维修大拿"], *PLA Daily*, August 17, 2019.

Chen Wengang [陈文刚] and Hou Biao [侯彪], "A Preliminary Inquiry into the Building of National Defense Mobilization Command Authorities in the Provincial Realm [省域国防动员指挥机构建设初探]," *National Defense* [国防], No. 7, 2019.

Cheng, Dean, "Converting the Potential to the Actual: Chinese Mobilization Policies and Planning," in Andrew Scobell, Arthur S. Ding, Phillip C. Saunders, and Scott W. Harold, eds., *The People's Liberation Army and Contingency Planning in China*, National Defense University Press, 2015.

Chieh, Chung, and Andrew N. D. Yang, "Crossing the Strait: Recent Trends in PLA 'Strategic Delivery' Capabilities," in Joel Wuthnow, Arthur Ding, Phillip C. Saunders, Andrew Scobell, and Andrew N. D. Yang, eds., *The PLA Beyond Borders: Chinese Military Operations in Regional and Global Context*, National Defense University Press, 2021.

Chieh, Chung, "PLA Logistics and Mobilization Capacity in a Taiwan Invasion," in Joel Wuthnow, Derek Grossman, Phillip C. Saunders, Andrew Scobell, and Andrew N. D. Yang, eds., *Crossing the Strait: China's Military Prepares for War with Taiwan*, National Defense University Press, 2022.

Cheung, Tai Ming, "Keeping Up with the *Jundui*: Reforming the Chinese Defense Acquisition, Technology, and Industrial System," in Philip C. Saunders, Arthur S. Ding, Andrew Scobell, Andrew N. D. Yang, and Joel Wuthnow, eds., *Chairman Xi Remakes the PLA: Assessing Chinese Military Reforms*, National Defense University Press, 2019.

China Air Force Encyclopedia Editorial Committee [中国空军百科全书编审委员会], *China Air Force Encyclopedia (Vol. 1)* [中国空军百科全书 (上卷)], Aviation Industry Press [航空工业出版社], 2005.

China Association for Quality [中国质量协会], ed., *On-Site Management Guide for Military Equipment Maintenance and Support Enterprises* [军用装备维修保障企业现场管理指南], China Standards Press [中国标准出版社], 2018.

China Global Television Network, "Now Live: 'Ordnance Master' Maintenance Platoon Relay Race" [正在直播: "军械能手"维修排接力赛], webpage, August 14, 2019. As of June 1, 2023:
http://www.mod.gov.cn/zt/2019-08/14/content_4848215.htm

"China Investigates Senior Military Logistics Officer," Reuters, October 21, 2015.

"Chinese Peacekeeping Medical Contingent to Mali Passes Pre-Mission Test," *China Military Online*, July 5, 2021.

"Circular of the State Council and the Central Military Commission on Issues Concerning Promoting the Socialization of Military Logistics Support" [国务院、中央军委关于推进军队后勤保障社会化有关问题的通知], Central People's Government of the People's Republic of China [中华人民共和国中央人民政府], June 14, 2018.

Clarke, John D., *Lifecycle Cost: An Examination of Its Application in the United States, and Its Potential for Use in the Australian Defense Forces*, thesis, Naval Postgraduate School, June 1990.

Clay, Marcus, and Dennis J. Blasko, "People Win Wars: The PLA Enlisted Force, and Other Related Matters," *War on the Rocks*, July 31, 2020.

Clay, Marcus, Dennis J. Blasko, and Roderick Lee, "People Win Wars: A 2022 Reality Check on PLA Enlisted Force and Related Matters," *War on the Rocks*, August 12, 2022.

Cliff, Roger, *China's Military Power: Assessing Current and Future Capabilities*, Cambridge University Press, 2015.

Congressional Budget Office, The All-Volunteer Force: Issues and Performance, 2007.

Correspondent Online [通信人在线], "Facilities Types and Equipment Types (GJB)" [设备及装备类(GJB)], webpage, January 12, 2020. As of May 31, 2023: http://www.txrzx.com/i5177.html

Defense Acquisition University, "Product Support Integrator (PSI) and Product Support Provider (PSP)," webpage, undated. As of June 1, 2023: https://www.dau.edu/acquipedia/pages/ArticleContent.aspx?itemid=395

Dillard, John, and Mark E. Nissen, *Determining the Best Loci of Knowledge, Responsibilities and Decision Rights in Major Acquisition Organizations*, Naval Postgraduate School, June 30, 2005.

"Eight 'Innovation Studios' Named After Officers and Soldiers" [8个"创新工作室"以官兵名字命名], *PLA Daily* [解放军报], March 2, 2020.

Engstrom, Jeffrey, *Systems Confrontation and System Destruction Warfare: How the Chinese People's Liberation Army Seeks to Wage Modern Warfare*, RAND Corporation, RR-1708-OSD, 2018. As of June 1, 2023: https://www.rand.org/pubs/research_reports/RR1708.html

Flanagan, Michael P., *Life Cycle Management Commands: Wartime Process or Long-Term Solution?* thesis, Army War College, 2007.

Fogarty, Gerard, "The Role of Organizational and Individual Differences Variables in Aircraft Maintenance Performance," *International Journal of Applied Aviation Studies*, Vol. 4, No. 3, March 2004.

Fowler, Eric S., *Culture and Military Effectiveness: How Societal Traits Influence Battle Outcomes*, thesis, Old Dominion University, Spring 2016.

Fravel, M. Taylor, *Active Defense: China's Military Strategy Since 1949*, Princeton University Press, 2019.

Friesendorf, Cornelius, *How Western Soldiers Fight: Organizational Routines in Multinational Missions*, Cambridge University Press, May 2018.

Gady, Franz-Stefan, "Revealed: Chinese Killer Drones in Iraq," *The Diplomat*, October 16, 2015.

Garafola, Cristina L., and Timothy R. Heath, *The Chinese Air Force's First Steps Toward Becoming an Expeditionary Air Force*, RAND Corporation, RR-2056-AF, 2017. As of June 1, 2023:
https://www.rand.org/pubs/research_reports/RR2056.html

Garafola, Cristina L., Timothy R. Heath, Christian Curriden, Meagan L. Smith, Derek Grossman, Nathan Chandler, and Stephen Watts, *The People's Liberation Army's Search for Overseas Basing and Access: A Framework to Assess Potential Host Nations*, RAND Corporation, RR-A1496-2, 2022. As of June 1, 2023:
https://www.rand.org/pubs/research_reports/RRA1496-2.html

Goldenberg, Irina, Manon Andres, Johan Österberg, Sylvia James-Yates, Eva Johansson, and Sean Pearce, "Integrated Defence Workforces: Challenges and Enablers of Military-Civilian Personnel Collaboration," *Journal of Military Studies*, Vol. 8, 2019.

Goldstein, Lyle J., "America Cannot Ignore China's Military Logistics Modernization," *National Interest*, October 11, 2021.

Gong Chuanxin [龚传信], ed., *Lectures on Military Equipment* [军事装备学教程], PLA Press [解放军出版社], 2004.

Gong Peiwen [巩沛文] and Tong Zujing [童祖静], "A Brigade of the 72nd Group Army Uses Self-Renovated Maintenance Equipment to Improve the Level of Battlefield Support" [第72集团军某旅运用自主革新的维修装备, 提升战场保障水平], *PLA Daily* [解放军报], September 13, 2021.

GSD—See General Staff Department.

Gunness, Kristen, "The Dawn of an Expeditionary PLA?" in Nadège Rolland, ed., *Securing the Belt and Road Initiative: China's Evolving Military Engagement Along the Silk Road*, National Bureau of Asian Research, September 2019.

Guo Kexin [郭克鑫] and Hai Yang [海洋], "A Brigade of the 79th Group Army Established an Equipment Maintenance Information Database to Facilitate Precision Support" [第79集团军某旅建立装备维修信息数据库助力精准保障], PLA Daily, May 23, 2021.

Han Qiulu [韩秋露], Kong Zhaojun [孔昭君], Deng Xiaotong [邓晓童], "Analysis of National Defense Mobilization Construction Under the Background of Military-Civilian Fusion" [军民融合背景下的国防动员建设分析], China Economic and Trade Herald [中国经贸导刊], No. 3, March 2019.

Hatzung, Scott A., and David B. Welborn, Leveraging Maintainer Experience to Increase Aviation Readiness, thesis, Naval Postgraduate School, 2020.

Henley, Lonnie, "PLA Logistics and Doctrine Reform, 1999-2009," in Susan M. Puska, ed., People's Liberation Army After Next, Strategic Studies Institute, August 2000.

Henley, Lonnie D., Civilian Shipping and Maritime Militia: The Logistics Backbone of a Taiwan Invasion, China Maritime Studies Institute, May 2022.

Hirrel, Leo P., "World War I and the Emergence of Modern Army Sustainment," in Keith R. Beurskens, ed., The Long Haul: Historical Case Studies of Sustainment in Large-Scale Combat Operations, Army University Press, 2018.

Hofstede, Geert, Culture's Consequences: International Differences in Work-Related Values, Sage Publications, 1984.

Hong Xuezhi, "The CPVF's Combat and Logistics," in Xiaobing Li, Allan R. Millett, and Bin Yu, eds., Mao's Generals Remember Korea, University Press of Kansas, 2001.

Huang, Kristin, "Chinese Military Short of Troops Trained in Hi-Tech Operations, PLA Daily Reveals in Rare Show of Candour," South China Morning Post, January 2, 2023.

Huang Shihai [黄世海], Armed Forces Equipment Management Outline [部队装备管理概论], Academy of Military Sciences [军事科学出版社], 2001.

"In 1985, the Wuhan Military Region Was Dismantled, and the Henan Province and Hubei Province Military Regions Within Its Jurisdiction; How to Transfer and Adjust?" [1985年，武汉军区撤编，辖内河南省、湖北省军区，如何转隶调整？], Cyclical Rate History [周期率历史], January 27, 2023.

"Information Technology Helps Improve Military Food Supply," China Military Online, March 15, 2021.

International Institute for Strategic Studies, The Military Balance 2023, 2023.

Jia Baohua [贾保华], Yang Lei [杨磊], and Xiang Shuangxi [相双喜], "Compiling an 'Encyclopedia' for Equipment Maintenance" [为装备维修编制"百科全书"] PLA Daily, April 5, 2018.

Jiang Tiejun [蒋铁军] and Zhang Huaiqiang [张怀强], *Ship Equipment Support Engineering Series: Equipment Maintenance Support Planning Management* [舰船装备保障工程丛书–装备维修保障计划管理], China Science Publishing [科学出版社], 2021.

"Jinan MR Infantry Division Relies on Local Support in Long-Range Maneuver," *PLA Daily Online*, October 24, 2010.

Joint Chiefs of Staff, *Joint Logistics*, Joint Publication 4-0, February 4, 2019, change 1, May 8, 2019.

Kane, Thomas, *Military Logistics and Strategic Performance*, Taylor & Francis, 2001.

Kania, Elsa B., and Lorand Laskai, "Myths and Realities of China's Military-Civil Fusion Strategy," Center for a New American Security, 2021

"Keep an Eye on Combat Training Support and Cultivating Scientific and Technological Talents" [紧盯战训保障 培育科技人才], *PLA Daily* [解放军报], June 7, 2022.

Kolleda, David J., *Force Protection Through Security of the Ground Lines of Communication (GLOC)*, U.S. Army Command and General Staff College, 1995.

Kraatz, Matthew S., "Learning by Association? Interorganizational Networks and Adaptation to Environmental Change," *Academy of Management Journal*, Vol. 41, No. 6, December 1998.

Lague, David, and Maryanne Murray, "T-Day: The Battle for Taiwan," Reuters, November 5, 2021.

Laksmana, Evan Abelard, *Imitation Game: Military Institutions and Westernization in Indonesia and Japan*, dissertation, Syracuse University, 2019.

Lawson, M. B., "In Praise of Slack: Time Is of the Essence," *Academy of Management Executive*, Vol. 15, No. 3, August 2001.

Lei Zhaoqiang [雷兆强], Wen Suyi [闻苏轶], and Li Peijin [李沛锦], "How to Face the New Challenges Brought by the New Position? This 'Ace Maintenance Worker' Gives the Answer" [如何面对新岗位带来的新挑战?这名"王牌维修工"给出答案], *PLA Daily*, April 9, 2021.

"Li Bei: Use 'X-Ray Eyes' to Detect the War Eagles' Flaws" [李蓓: 用"透视眼"为战鹰探伤], *PLA Daily*, October 25, 2018.

Li Tuanbiao [李团标], Xing Zhe [邢哲], and Lei Zhaoqiang [雷兆强], "An NCO and His 400 Plus Issues of Maintenance Tabloids" [一名士官和他的400多期维修小报], PLA Daily, December 23, 2019.

Li Weixin [李伟欣], Wei Yumeng [卫雨檬], Shu Xiquan [殳细泉], and Wu Lihua [吴李华], "'This Is My Fighter Jet!' Salute! Air Force Mechanic" ["这是我的战机!" 致敬! 空军机务兵], *PLA Daily* [解放军报], January 7, 2021.

Light, Thomas, Daniel M. Romano, Michael Kennedy, Caolionn O'Connell, and Sean Bednarz, *Consolidating Air Force Maintenance Occupational Specialties,* RAND Corporation, RR-A1307-AF, 2016. As of June 1, 2023: https://www.rand.org/pubs/research_reports/RR1307.html

Lin, Bonny, and Cristina L. Garafola, *Training the People's Liberation Army Air Force Surface-to-Air Missile (SAM) Forces,* RAND Corporation, RR-1414-AF, 2016. As of June 1, 2023: https://www.rand.org/pubs/research_reports/RR1414.html

Liu Hanbao [刘汉宝], "Approaching 4 Soldiers 'Craftsmen' in Equipment Maintenance Posts: Dedication and Perseverance in the Rear Area of the Battlefield" [走近4名装备维修岗位的士兵"工匠":战场后方的执着与坚守], *PLA Daily,* April 12, 2022.

Liu Tielin [刘铁林], Wu Yongle [武永乐], Zhang Haifeng [张海峰], and Huang Xinxin [黄欣鑫], *Army Unit-Level Equipment Maintenance Support Operation Mode* [陆军部队级装备维修保障作业模式], National Defense Industry Press [国防工业出版社], 2022.

Liu Wenkai [刘文开], Wang Hui [王晖], Ma Yali [马雅丽], He Peng [何鹏], and Chen Shaoshan [陈绍山], "Overall Conception of Theater Army Equipment Support Classification" [战区陆军装备保障体系建设总体构想], *Journal of the Academy of Armored Force Engineering* [装甲兵工程学院学报], Vol. 32, No. 6, 2018.

Liu Xuanzun, "Over 200 Chinese CH-4 Drones Sold on International Market; New Facility to Expand Production Capacity: Manufacturer," *Global Times,* November 2, 2022.

Liu Yibo and Chen Pengfei, "Inflatable Camping Tents Distributed to Plateau Troops," *China Military Online,* May 7, 2021.

Lu Ke [卢科] and Sun Xingwei [孙兴维], "The Joint Logistic Support Force Innovates with Three Mechanisms to Comprehensively Upgrade the Quality of General Engineering Construction" [联勤保障部队创新3项机制 全面提升通用工程建设质量], *PLA Daily* [解放军报], June 4, 2022.

Luce, LeighAnn, and Erin Richter, "Handling Logistics in A Reformed PLA: The Long March Toward Joint Logistics," in Phillip C. Saunders, Arthur S. Ding, Andrew Scobell, Andrew N. D. Yang, and Joel Wuthnow, eds., *Chairman Xi Remakes the PLA: Assessing Chinese Military Reforms*, National Defense University Press, 2019.

Luo Yi [罗祎], Su Zhiyang [苏执阳], Ruan Minzhi [阮旻智], and Liu Tianhua [刘天华], *Military-Use Equipment Maintenance Support Resources Prediction and Deployment Technology* [军用装备维修保障资源预测与配置技术], Weapons Industry Press [兵器工业出版社], 2015.

Maywald, Jacob, Benjamin Hazen, Edward Salo, and Michael Hugos, "Logistics Interdiction for Taiwan Unification Campaigns," *War on the Rocks*, August 21, 2023.

McCauley, Kevin, written testimony for hearing on China's Military Power Projection and U.S. National Interests, U.S.-China Economic and Security Review Commission, February 20, 2020.

McCauley, Kevin, *Logistics Support for a Cross-Strait Invasion: The View from Beijing*, China Maritime Studies Institute, July 2022.

McGrath, John J., *The Other End of the Spear: The Tooth-to-Tail Ratio (T3R) in Modern Military Operations*, Combat Studies Institute Press, 2007.

Military Leak, "CASC to Expand Production Capacity of CH-4 Rainbow Unmanned Aerial Vehicles," November 7, 2022.

Millet, Allan R., Williamson Murray, and Kenneth H. Watman, "The Effectiveness of Military Organizations," in Allan R. Millet and Williamson Murray, eds., *Military Effectiveness, Volume 1, The First World War*, Cambridge University Press, 1988.

Min, Ryan B., *Limits of Decentralization: Streamlining the Dispersed Parts of the Information Whole*, thesis, U.S. Army Command and General Staff College, 2018.

"The Ministry of National Defense Introduced the Construction and Development of the Joint Logistic Support Force Since Its Reform and Reconstruction" [国防部介绍联勤保障部队改革重塑以来建设发展情况], Ministry of National Defense Network [国防部网], September 30, 2021.

Mitzer, Stijn, "Operational Failure: The CH-4B's Short-Lived Career in Jordan," *Oryx* blog, January 17, 2022.

Mitzer, Stijn, and Joost Oliemans, "Tracking Worldwide Losses of Chinese-Made UAVs," *Oryx* blog, November 19, 2021.

Mulvenon, James, "PLA Divestiture 2.0: We Mean It This Time," *China Leadership Monitor*, No. 50, July 9, 2016.

"National Defense Mobilization Law of the PRC (Full Text)" [中华人民共和国国防动员法 (全文)], Xinhua [新华网], February 26, 2010.

Ng, Jr, "China's CASC Wins Follow-On CH-4 UAV Export Order," *Asian Military Review*, September 16, 2022.

Nguyen, Hieu Thanh, Kieu Trang Vu, Loan Quynh Thi Nguyen, and Hiep Ngoc Luu, "CEO Culture, Corruption, and Firm Performance," *Applied Economics Letters*, Vol. 29, No. 7, April 2022.

Nong Qinghua [农清华], "Reform of PLA Logistics Support System in the Past 40 Years" [人民解放军后勤保障体制改革攻坚40年], *Military History* [军事历史], No. 1, 2019.

O'Dowd, Edward C., *Chinese Military Strategy in the Third Indochina War: The Last Maoist War*, Routledge, 2007.

Office of the Secretary of Defense, *Annual Report to Congress: Military and Security Developments Involving the People's Republic of China 2019*, Department of Defense, May 2019.

Office of the Secretary of Defense, *Annual Report to Congress: Military and Security Developments Involving the People's Republic of China*, Department of Defense, 2022.

Oliver, Amalya L., "Strategic Alliances and the Learning Life-Cycle of Biotechnology Firms," *Organization Studies,* Vol. 3, May 2001.

Paoli, Massimo, and Andrea Prencipe, "Memory of the Organisation and Memories Within the Organisation," *Journal of Management and Governance*, Vol. 7, 2003.

Patankar, Manoj S., "A Study of Safety Culture at an Aviation Organization," *International Journal of Applied Aviation Studies*, Vol. 3, No. 1, March 2003.

Peng Guangqian [彭光谦] and Yao Youzhi [姚有志], eds., *Science of Military Strategy* [战略学], Academy of Military Science Press [军事科学出版社], 2001.

People's Liberation Army General Staff Department Military Training and Arms, *Research into the Afghanistan War*, Liberation Army Press, 2004.

People's Liberation Army General Staff Department Military Training Department, *Research into the Kosovo War*, Liberation Army Press, 2000.

Pettersen, Kenneth A., and Karina Aase, "Explaining Safe Work Practices in Aviation Line Maintenance," *Safety Science,* Vol. 46, No. 3, March 2008.

Pettyjohn, Stacie, Becca Wasser, and Chris Dougherty, *Dangerous Straits: Wargaming a Future Conflict over Taiwan*, Center for a New American Security, June 2022.

Pierce, William G., *Span of Control and the Operational Commander: Is It More Than Just a Number?* U.S. Army Command and General Staff College, 1991.

Powell, James S., *Taking a Look Under the Hood: The October War and What Maintenance Approaches Reveal about Military Operations*, Institute of Land Warfare, Association of the United States Army, August 2019.

Puska, Susan M., "Taming the Hydra: Trends in China's Military Logistics Since 2000," in Roy Kamphausen, David Lai, and Andrew Scobell, eds., *The PLA at Home and Abroad: Assessing the Operational Capabilities of China's Military*, Strategic Studies Institute, 2010.

Puska, Susan M., Aaron Shraberg, Daniel Alderman, and Jana Allen, "A Model for Analysis of China's Defense Life Cycle Management System," *Study of Innovation and Technology in China Policy Briefs*, No. 4, 2014.

"Report: PLA Navy Runs into Crewing Difficulties for Growing Fleet," *Maritime Executive*, January 3, 2023.

Richter, Erin, Leigh Ann Ragland, and Katherine Atha, "General Logistics Department Organizational Reforms: 2000-2012" in Kevin Pollpeter and Kenneth W. Allen, eds., *The PLA as Organization v2.0*, Defense Group Inc., 2012.

Ritschel, Jonathan D., and Tamiko L. Ritschel, "Organic or Contract Support? Investigating Cost and Performance in Aircraft Sustainment," *Journal of Transportation Management*, Vol. 26, No. 2, Fall–Winter 2016.

Roberts, Karlene H., "Managing High Reliability Organizations," *California Management Review*, Vol. 32, No. 4, July 1990.

Rochlin, Gene I., Todd R. La Porte, and Karlene H. Roberts, "The Self-Designing High-Reliability Organization: Aircraft Carrier Flight Operations at Sea," *Naval War College Review*, Vol. 40, No. 4, Autumn 1987.

Ruan Yongjun [阮拥军] and Wang Lihong [王利洪], "On the Application of 3D Printing Technologies in Equipment Maintenance Support in Our Military's United Nations Peacekeeping Operations" [试述3D打印技术在我军联合国维和行动装备维修保障中的运用], *National Defense* [国防], No. 8, 2019.

Sagan, Scott D., *The Limits of Safety: Organizations, Accidents, and Nuclear Weapons*, Princeton University Press, 1993.

Shatzer, George R., and Roger D. Cliff, eds., *PLA Logistics and Sustainment: PLA Conference 2022*, U.S. Army War College Press, 2023.

Shrader, Charles R., *U.S. Military Logistics, 1607–1991: A Research Guide*, Greenwood, 1992.

Shi Feng [石峰] and Guo Rong [郭蓉], "Repair Is Like Playing Hide and Seek with Faults, but He Has 'Piercing Eyes'" [维修就像是与故障玩捉迷藏，可他有"火眼金睛"], *PLA Daily*, April 17, 2020.

Shou Xiaosong [寿晓松], ed., *Science of Military Strategy* [战略学], Academy of Military Science Press [军事科学院], 2013.

Shu Zhengping [舒正平], ed., *Science of Military Equipment Maintenance Support* [军事装备维修保障学], National Defense Industry Press [国防工业出版社], 2013.

Snyder, Don, Kristin F. Lynch, Colby Peyton Steiner, John G. Drew, Myron Hura, Miriam E. Marlier, and Theo Milonopoulos, *Command and Control of U.S. Air Force Combat Support in a High-End Fight*, RAND Corporation, RR-A316-1, 2021. As of June 1, 2023:
https://www.rand.org/pubs/research_reports/RRA316-1.html

Song Naihong [宋乃宏], "Some Thoughts on Doing Well in the Investigative and Statistical Work on National Defense Mobilization Potential in the New Era" [新时代做好国防动员潜力统计调查工作的几点思考], *National Defense* [国防], No. 8, 2019.

Stockholm International Peace Research Institute, "Trade Registers," webpage, undated. As of June 9, 2023:
https://armstrade.sipri.org/armstrade/page/trade_register.php

Sun Xingwei and Zhang Liang, "Chinese Military Organizes Special Production Training for Peacekeepers for First Time," *China Military Online*, June 21, 2021.

Sun Xingwei [孙兴维] and Zhang Peng [张鹏], "The Logistic Support Department of the Central Military Commission Issued the 'Military Personnel Health Examination Measures,'" [中央军委后勤保障部印发"军队人员健康体检办法"], Xinhua [新华社], April 27, 2022.

Sun Xiude [孙秀德] and Xu Qi [徐起], "The Direction of Military Logistic Support System Reforms [军队后勤保障体制改革的方向], *Military Economic Research* [军事经济研究], Vol. 7, 1998.

Sun Yuerui [孙月瑞] and Bai Lanxin [白兰鑫], "A Certain Engineering Brigade of the Air Force Organizes Actual Combat Training and Assessment to Temper the Battlefield Awareness of Officers and Soldiers" [空军某工兵大队组织实战化训练考核锤炼官兵战场意识], *PLA Daily* [解放军报], August 21, 2022.

Tagarev, Todor, "Enabling Factors and Effects of Corruption in the Defense Sector," *Connections*, Vol. 9, No. 3, Summer 2010.

Tang Wenyuan [汤文元] and Wu Shike [吴世科], "Li Xiangnan: 20 Years Maintaining Warhawks over a Thousand Times with Zero Mistakes" [李向楠: 20年维修战鹰千余次零失误], *PLA Daily*, January 17, 2019.

Tang Zhiyong [唐志勇] and Chen Wanjin [陈万金], "Retired Veteran Xu Daming: Repairing Boats for 30 Years; Ensured Nearly 1,000 Sailings with Zero Mistakes" [退役老兵徐达明:维修船艇30年, 近千次保障出航零失误], *PLA Daily*, December 8, 2020.

Terzi, Sergio, Abdelaziz Bouras, Debashi Dutta, Marco Garetti, and Dimitris Kiritsis, "Product Lifecycle Management—From Its History to Its New Role," *International Journal of Product Lifecycle Management*, Vol. 4, No. 4, November 2010.

"The 'Three Questions' of Professional Training in the Service Support Battalion" [勤务保障营的专业训练 "三问"], *PLA Daily* [解放军报], April 28, 2020.

Tkacik, John J., Jr., "From Surprise to Stalemate: What the People's Liberation Army Learned from the Korean War—A Half-Century Later," in Laurie Burkitt, Andrew Scobell, and Larry M. Wortzel, eds., *The Lessons of History: The Chinese People's Liberation Army at 75*, Strategic Studies Institute, 2003.

Trapans, Andris, *Logistics in Recent Soviet Military Writing*, RAND Corporation, RM-5602-PR, 1966. As of October 2, 2023:
https://www.rand.org/pubs/research_memoranda/RM5062.html

Trapans, Andris, *Organizational Maintenance in the Soviet Air Force*, RAND Corporation, RM-4382-PR, 1965. As of October 2, 2023:
https://www.rand.org/pubs/research_memoranda/RM4382.html

Tripp, Robert S., John G. Drew, and Kristin F. Lynch, *A Conceptual Framework for More Effectively Integrating Combat Support Capabilities and Constraints into Contingency Planning and Execution*, RAND Corporation, RR-1025-AF, 2015. As of June 1, 2023:
https://www.rand.org/pubs/research_reports/RR1025.html

U.S. Department of Defense Inspector General, "Operation Inherent Resolve: Lead Inspector General Report to the United States Congress," August 2, 2019.

U.S. Department of Defense Inspector General, "Operation Inherent Resolve: Lead Inspector General Report to the United States Congress," August 3, 2021.

U.S. Department of Defense Inspector General, "Operation Inherent Resolve: Lead Inspector General Report to the United States Congress," November 1, 2022.

U.S. Government Accountability Office, *F-35 Sustainment: DOD Needs to Cut Billions in Estimated Costs to Achieve Affordability*, July 2021.

Uzochukwu, Benedict, Silvanus Udoka, Paul Stanfield, and Eui Park, "Design for Sustainment—A Conceptual Framework," *Proceedings of the 2010 Industrial Engineering Research Conference*, January 2010.

Vershinin, Alex, "Feeding the Bear: A Closer Look at Russian Army Logistics and the Fait Accompli," *War on the Rocks*, November 23, 2021.

Voss, Nathan, and James Ryseff, *Comparing the Organizational Cultures of the Department of Defense and Silicon Valley*, RAND Corporation, RR-A1498-2, 2022. As of June 2, 2023:
https://www.rand.org/pubs/research_reports/RRA1498-2.html

Wang Fengcai [王凤才], "The Logistical Problems Russia Exposed in the Russia-Ukraine Conflict" [俄乌冲突中俄军战场后勤保障暴露的问题], *World Military Review* [外国军事学术], No. 5, 2022.

Wang Jizhen [王纪震] and Ding Guandong [丁冠东], "Equipment Mobilization Information System Construction" [装备动员信息系统构建], *Command Information System and Technology* [指挥信息系统与技术], Vol. 9, No. 3, 2018.

Wang Yi [王轶] and Cheng Jun [程俊], "A Brigade of the 82nd Group Army Innovated the Optimal Support Model Modular Grouping Helps Infantry Fighting Vehicles (IFV) Gallop" [第八十二集团军某旅创新集优保障模式 模块编组助力战车驰骋], *PLA Daily,* April 2, 2022.

Wang Yongming, Liu Xiaoli, and Xiao Yunhua, eds., *Research into the Iraq War,* Military Science Press, 2003.

Watkins, Emma, and Thomas Spoehr, *The Defense Production Act: An Important National Security Tool, But It Requires Work*, The Heritage Foundation, October 15, 2019.

Wentz, Larry, *An ICT Primer: Information and Communication Technologies for Civil-Military Coordination in Disaster Relief and Stabilization and Reconstruction*, Center for Technology and National Security Policy, National Defense University, July 2006.

"Why Does Our Military's J-10 Have Two Production Batch Numbers and Parts That Come from Three Fighters?" [我军这架歼10为何有两个生产批号 零件来自3架战机], *Sina Military* [新浪军事], January 11, 2019.

Wong, Kelvin, "Update: CASC Expands Domestic UAV Testing and Production Capabilities," *Janes*, January 8, 2022.

Wuthnow, Joel, "Responding to the Epidemic in Wuhan: Insights into Chinese Military Logistics," *China Brief,* Vol. 20, No. 7, April 13, 2020.

Wuthnow, Joel, "A New Era for Chinese Military Logistics," *Asian Security*, Vol. 17, No. 3, September–December 2021.

Wuthnow, Joel, "Joint Logistic Force Support to Theater Commands," *PLA Logistics and Sustainment: PLA Conference 2022*, U.S. Army War College Press, 2023.

Wuthnow, Joel, Arthur S. Ding, Phillip C. Saunders, Andrew Scobell, and Andrew N. D. Yang, eds., *The PLA Beyond Borders: Chinese Military Operations in Regional and Global Context*, National Defense University Press, 2021.

Wuthnow, Joel, and Phillip C. Saunders, *Chinese Military Reforms in the Age of Xi Jinping: Drivers, Challenges, and Implications*, Institute for National Strategic Studies, National Defense University, March 2017.

Wuthnow, Joel, Phillip C. Saunders, and Ian Burns McCaslin, "PLA Overseas Operations in 2035: Inching Toward a Global Combat Capability," *Strategic Forum*, No. 309, National Defense University Press, May 2021.

Xia Dong [夏冬] and Yang Fan [杨帆], "Air Force Engineering University Aviation Non-Commissioned Officer School Lifts the 'Wings' of the War Eagle to Take Off" [空军工程大学航空机务士官学校托举战鹰腾飞的"翅膀"], Xinhua, April 25, 2021.

Xiao Tianliang [肖天亮], ed., *Science of Military Strategy* [战略学], National Defense University Publishing House [国防大学出版社], 2017.

Xiao Tianliang [肖天亮], ed., *Science of Military Strategy* [战略学], National Defense University Press [国防大学出版社], 2020.

Xiaobing Li, *A History of the Modern Chinese Army*, University Press of Kentucky, 2007.

Xiong Chencheng [熊晨成] and Hu Ansheng [胡安胜], "Research on Coordinates-Based ABC Classification Management of Loaded Materiel" [基于坐标系的被装物资 ABC 分类管理研究], *Logistics Technology* [物流技术], No. 12, 2018.

Xu Jinzhang [胥金章] and Zhang Yuqing [张玉清], "With the Approval of the Central Military Commission, the Jinan Theater Has Carried Out a Pilot Program of Large-Scale Joint Logistics Reform" [经中央军委批准 济南战区进行大联勤改革试点], Xinhua, June 25, 2004.

Xu Xu [徐徐], Sun Qilong [孙启龙], and Xing Zhe [邢哲], "From One Person to a Team, Explore the 'Correct Way to Open' Equipment Maintenance Support" [从一个人到一个团队，探寻装备维修保障的"正确打开方式"]," *PLA Daily*, May 28, 2021.

Yang Fan [杨帆] and Li Jianwen [李建文], "Instructors from the Air Force Engineering University Aviation Non-Commissioned Officer School Arrive with the Troops to Follow the Flight Practice" [空军工程大学航空机务士官学校教员到部队跟飞实践], *PLA Daily*, August 30, 2020.

Yang Pan [杨盼], "Cold Training: What to Do If the War Eagle Is Cold? The Air Force Pilots Who Participated in the Cold Training Will Tell You" [寒训 | 战鹰冷了怎么办?参加寒训的空军机务兵告诉你], *China Military Online* [中国军网], January 4, 2019.

Yang Xueming [杨学铭] and Xun Ye [荀烨], "Study on Theater Ground Force Materiel Distribution and Support Network Mode Under the New System" [新体制下战区陆军物资配送保障网络模式研究], *Logistics Technology* [物流技术], No. 6, 2018.

Yin Weihua [尹威华] and Zhou Huanquan [周焕权], "Expert Plateau Soldiers Have Maintenance 'Knacks' for Emergency Repair of Vehicles, Becoming a 'Reassuring Pill' for Fellow Soldiers" [高原兵专家维修有"诀抢修车成了战友的"定心丸"], Ministry of National Defense Network, August 28, 2018.

Yoshihara, Toshi, *Chinese Lessons from the Pacific War: Implications for PLA Warfighting*, Center for Strategic and Budgetary Assessments, January 5, 2023.

Zeng Tingze [曾廷泽] and Pan Dahong [潘大红], "The Enlightenment of Logistic Support from Kosovo War" [从科索沃战争看未来后勤保障], *Military Economic Research* [军事经济研究], 2000.

Zhang Hongbo [张洪波] and Lei Ziyuan [雷梓园], "Breaking Through the 'Ceiling' of Technology, Feng Jianzhong Jumped from an Ordinary Soldier to a Maintenance Expert" [冲开技术的"天花板",冯建中从普通一兵跃升为维修专家], *PLA Daily*, June 27, 2019.

Zhang Taiyao [张泰耀] and Huang Yongjian [黄永健], "Troubleshooting Introduces 'Cloud Maintenance'" [故障处置引入"云端维修"], *PLA Daily*, October 30, 2022.

Zhang Yuqing [张玉清], Zhou Zhengxin [周正信], and Yang Fan [杨帆], "Air Force Engineering University Aviation Non-Commissioned Officer School Organizes Teachers to Go Fly with the Troops" [空军工程大学航空机务士官学校组织教员赴部队跟飞学习], Xinhua, August 15, 2019.

Zhang Yuqing [张玉清], Zhang Lei [张雷], Feng Bin [冯斌], "Record of Bai Wenguo, a Mechanical Technician in the Second Maintenance Squadron of the Air Force's August 1st Acrobatic Team" [擦亮名片扬国威—记空军八一飞行表演队机务二中队机械技师白文国], Xinhua, June 5, 2019.

Zhang Zhaoyang [张照洋] and Huang Kainan [黄凯楠], "A Brigade of the 72nd Group Army: Practicing New Equipment and Strengthening Old Equipment" [第72集团军某旅：练熟新装备 练强老装备], *PLA Daily*, April 23, 2022.

Zhao Jianwei [赵建伟], "China's Military Logistics Reform Plan Is Clear, and Substantial Progress Has Been Made in Four Aspects" [中国军事后勤变革图渐清晰 四方面获实质性进展], *ChinaNews* [中国新闻网], April 24, 2007.

Zhao Lusheng [赵潞生], ed., *Science of Military Equipment* [军事装备学], National Defense University Press [国防大学出版社], 2000.

Zheng Dongliang [郑东良], ed., *Aviation Maintenance Theory* [航空维修理论], National Defense Industry Press [国防工业出版社], 2012.